Fuel Gas Code
of New York State

**New York State
Department of State**
Division of Code Enforcement and Administration

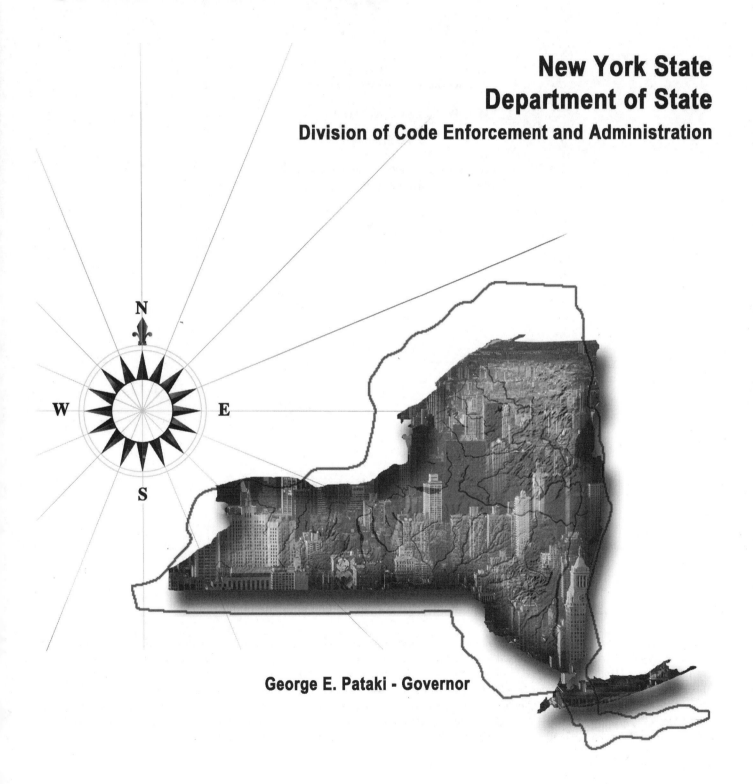

George E. Pataki - Governor

Fuel Gas Code of New York State

Publication Date: May 2002

First Printing

ISBN # 1-58001-090-3 (soft-cover edition)
ISBN # 1-58001-094-6 (loose-leaf edition)

ACKNOWLEDGEMENTS

The Department of State gratefully acknowledges the following individuals who contributed to the development of the Plumbing, Mechanical and Fuel Gas Codes of New York State:

State Fire Prevention and Building Code Council

Randy A. Daniels, Secretary of State (Chair)
Alexander F. Treadwell, former Secretary of State (Chair 1995 - 2001)
John W. Hasper, Deputy Secretary of State (designee)
James A. Burns, State Fire Administrator
Ogden J. Clark (designee)
Antonia Coello Novello, M.D., Commissioner of Health
Barbara DeBuono, M..D., former Commissioner of Health
Richard Svenson (designee)
Linda Angello, Commissioner of Labor
James McGowan, former Commissioner of Labor
Denis Peterson (designee)
Thomas V. Ognibene, Councilman, City of New York
Roy A. Bernardi, former Mayor, City of Syracuse
Nick Altieri (designee)
James P. Griffin, Mayor, City of Olean
Christopher Young (designee)
Michael Behling, Legislator, County of Jefferson
Paul Noto, Legislator, County of Westchester
Kevin Donohue, Councilman, Town of LaGrange
Scott Wohl, Trustee, Village of Goshen
Stephen Brescia, Mayor, Village of Montgomery
Carmen Dubaldi (designee)
Gunnar Neilson, Fire Service Official
Robert G. Shibley, AIA, Registered Architect
Ronald Bugaj, Registered Architect (deceased)
Dr. James J. Yarmus, P.E., Professional Engineer
John H. Flanigan, Code Enforcement Official
Robert Hankin, Builders' representative
John J. Torpey, Trade union representative
Terence J. Moakley, Persons with disabilities representative

Department of State

Frank Milano, First Deputy Secretary of State
Dorothy M. Harris, Assistant Secretary of State (Project coordinator)
George E. Clark, Jr., Director, Division of Code Enforcement and Administration
Michael Saafir, Deputy Director
Ronald Piester, Assistant Director of Code Development
Richard DiGiovanna, Office of Counsel

Plumbing, Mechanical and Fuel Gas Codes Technical Subcommittee

Cheryl Fischer (Chair), John Addario, Michael Burnetter, Bill Richardsen, Kumar Vijaykumar (Department of State Staff)
Lou Ackerman, Robert Cordell, Marsha Dollendorf, Dr. Larry Feeser, Jim Hart, Ron Monast, Michael Montysko

Administrative Task Group

Richard Thomson (Chair), James King, Thomas Romanowski, William Stewart, Robert Thompson

International Conference of Building Officials - Publications Staff

Kim Akhavan (Managing Editor), Alberto Herrera (Typesetter), Yolanda Nickoley (Typesetter), Suzane Nunes (Product Development Manager), Mary Bridges, Marje Cates, Sally Clem, Greg Dickson, Carmel Gieson, Jessica Hoffman, Roger Mensink, Rhonda Moller, Scott Pierce, Cindy Rodriguez, Mike Tamai, Lisa Valentino, Brian Wohn

Cover Photograph

Dave Feiden

Content

The *Fuel Gas Code of New York State* combines language from the 2000 *International Fuel Gas Code*®, 2001 Supplement to the *International Fuel Gas Code*, and New York modifications developed by the State Fire Prevention and Building Code Council and its Plumbing, Mechanical and Fuel Gas Code Technical Subcommittee. In addition, administrative modifications to the 2000 *International Fuel Gas Code* were developed by the Department of State's Administrative Task Group.

Marginal Markings

New York modifications to code language are indicated by NY tape (⸸) in the margin, and New York text is underlined. Deletion of code language by New York is indicated by an arrow (⇒) in the margin.

"Reserved" indicates that a section or portion of the International Code™ has been deleted, but its number or position has been retained.

Letter Designations in Front of Section Numbers

The content of sections in this code which begin with a letter designation are maintained by another code development committee in accordance with the following: [B] = International Building Code Development Committee; [F] = International Fire Code Development Committee; [M] = International Mechanical Code Development Committee; [P] = International Plumbing Code Development Committee; [RBE] = International Residential Code Building and Energy Development Committee; [RMP] = International Residential Code Mechanical/Plumbing Development Committee; and [E] = International Energy Conservation Code Development Committee.

TABLE OF CONTENTS

CHAPTER 1   GENERAL REQUIREMENTS 1

Section

101   General 1
102   Applicability 1
103   Office of Code Enforcement 2
104   Code Enforcement Official 2
105   Approval 2
106   Permits 3
107   Inspections 3
108   Violations 3
109   Variance Procedures 3

CHAPTER 2   DEFINITIONS 5

Section

201   General 5
202   General Definitions 5

CHAPTER 3   GENERAL REGULATIONS 15

Section

301   General 15
302   Structural Safety 16
303   Appliance Location 18
304   Combustion, Ventilation, and
           Dilution Air 19
305   Installation 21
306   Access and Service Space 22
307   Condensate Disposal 23
308   Clearance Reduction 23
309   Electrical 24

CHAPTER 4   GAS PIPING INSTALLATIONS . . . 27

Section

401   General 27
402   Pipe Sizing 27
403   Piping Materials 46
404   Piping System Installation 48
405   Pipe Bends and Changes in
           Direction 50
406   Inspection, Testing, and
           Purging 50
407   Piping Support 52
408   Drips and Sloped Piping 52
409   Shutoff Valves 53
410   Flow Controls 53
411   Appliance Connections 54
412   Liquefied Petroleum Gas Motor Vehicle
           Fuel-Dispensing Stations 54
413   Compressed Natural Gas Motor Vehicle
           Fuel-Dispensing Stations 55

414   Supplemental and Standby
           Gas Supply 57
415   Piping Support Intervals 57

CHAPTER 5 CHIMNEYS AND VENTS 59

Section

501   General 59
502   Vents 60
503   Venting of Equipment 61
504   Sizing of Category I Appliance
           Venting Systems 72
505   Direct Vent, Integral Vent, Mechanical
           Vent and Ventilation/Exhaust
           Hood Venting 76
506   Factory-Built Chimneys 76

CHAPTER 6   SPECIFIC APPLIANCES 93

Section

601   General 93
602   Decorative Appliances for
           Installation in Fireplaces 93
603   Log Lighters 93
604   Vented Gas Fireplaces (Decorative Appliances)
           and Vented Gas Fireplace Heaters 93
605   Incinerators and Crematories 93
606   Commercial-Industrial
           Incinerators 93
607   Vented Wall Furnaces 93
608   Floor Furnaces 94
609   Duct Furnaces 94
610   Direct-Fired Make-Up
           Air Heaters 95
611   Direct-Fired Industrial
           Air Heaters 95
612   Clothes Dryers 96
613   Clothes Dryer Exhaust 96
614   Sauna Heaters 96
615   Engine and Gas Turbine-Powered
           Equipment 97
616   Pool and Spa Heaters 97
617   Forced-Air Warm-Air
           Furnaces 97
618   Conversion Burners 98
619   Unit Heaters 98
620   Unvented Room Heaters 99
621   Vented Room Heaters 99
622   Cooking Appliances 99
623   Water Heaters 99
624   Refrigerators 100
625   Gas-Fired Toilets 100
626   Air-Conditioning
           Equipment 100

627   Illuminating Appliances 101
628   Small Ceramic Kilns 101
629   Infrared Radiant Heaters 101
630   Boilers 101
631   Equipment Installed in Existing
       Unlisted Boilers 102
632   Chimney Damper Opening
       Area 102
633   Fuel Cell Power Plants 102

CHAPTER 7   **REFERENCED STANDARDS** 103

APPENDIX A   **SIZING AND CAPACITIES OF**
                    GAS PIPING 111

APPENDIX B   **SIZING OF VENTING SYSTEMS**
                    SERVING APPLIANCES
                    EQUIPPED WITH DRAFT
                    HOODS, CATEGORY 1
                    APPLIANCES, AND
                    APPLIANCES LISTED
                    FOR USE AND TYPE B
                    VENTS 115

APPENDIX C 125

APPENDIX D   **RECOMMENDED PROCEDURE**
                    FOR SAFETY INSPECTION OF
                    AN EXISTING APPLIANCE
                    INSTALLATION 127

APPENDIX E   **STRUCTURAL SAFETY** 129

INDEX 131

SECTION 101
GENERAL

101.1 Title. These provisions shall be known as the *Fuel Gas Code of New York State* and shall be cited as such, and will be referred to herein as "this code."

101.2 Scope. The provisions of this code shall apply to the installation of fuel gas piping systems, fuel gas utilization equipment, and fuel related accessories as follows:

1. Coverage of piping systems shall extend from the point of delivery to the connections with gas utilization equipment. (See "point of delivery.")

2. Systems with an operating pressure of 125 psig (862 kPa gauge) or less.

   Piping systems for gas-air mixtures within the flammable range with an operating pressure of 10 psig (69 kPa gauge) or less.

   LP-Gas piping systems with an operating pressure of 20 psig (140 kPa gauge) or less.

3. Piping systems requirements shall include design, materials, components, fabrication, assembly, installation, testing, inspection, operation, and maintenance.

4. Requirements for gas utilization equipment and related accessories shall include installation, combustion, and ventilation air and venting.

This code shall not apply to the following:

1. Portable LP-Gas equipment of all types that are not connected to a fixed fuel piping system.

2. Installation of farm equipment such as brooders, dehydrators, dryers, and irrigation equipment.

3. Raw material (feedstock) applications except for piping to special atmosphere generators.

4. Oxygen-fuel gas cutting and welding systems.

5. Industrial gas applications using gases such as acetylene and acetylenic compounds, hydrogen, ammonia, carbon monoxide, oxygen, and nitrogen.

6. Petroleum refineries, pipeline compressor or pumping stations, loading terminals, compounding plants, refinery tank farms, and natural gas processing plants.

7. Integrated chemical plants or portions of such plants where flammable or combustible liquids or gases are produced by chemical reactions or used in chemical reactions.

8. LP-Gas installations at utility gas plants.

9. Liquefied natural gas (LNG) installations.

10. Fuel gas piping in power and atomic energy plants.

11. Proprietary items of equipment, apparatus, or instruments such as gas generating sets, compressors, and calorimeters.

12. LP-Gas equipment for vaporization, gas mixing, and gas manufacturing.

13. Temporary LP-Gas piping for buildings under construction or renovation that is not to become part of the permanent piping system.

14. Installation of LP-Gas systems for railroad switch heating.

15. Installation of LP-Gas and compressed natural gas (CNG) systems on vehicles.

16. Gas piping, meters, gas pressure regulators, and other appurtenances used by the serving gas supplier in the distribution of gas, other than undiluted LP-Gas.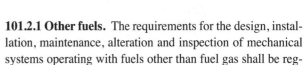

101.2.1 Other fuels. The requirements for the design, installation, maintenance, alteration and inspection of mechanical systems operating with fuels other than fuel gas shall be regulated by the *Mechanical Code of New York State*.

101.3 Appendices. The following appendix has been adopted and is made part of this code:

Appendix E - Structural Safety

101.4 Purpose. The purpose of this code is to provide minimum standards to safeguard life or limb, health, property and public welfare by regulating and controlling the design, construction, installation, quality of materials, location, operation and maintenance or use of fuel gas systems.

101.5 Severability. If a section, subsection, sentence, clause or phrase of this code is, for any reason, held to be unconstitutional, such decision shall not affect the validity of the remaining portions of this code.

SECTION 102
APPLICABILITY

102.1 General. The provisions of this code shall apply to all matters affecting or relating to structures and premises, as set forth in Section 101. Where, in a specific case, different sections of this code specify different materials, methods of construction or other requirements, the most restrictive shall govern.

102.2 Existing installations. Except as otherwise provided for in this chapter, a provision in this code shall not require the removal, alteration or abandonment of, nor prevent the continued utilization and maintenance of, existing installations lawfully in existence at the time of the adoption of this code.

102.3 Maintenance. Installations, both existing and new, and parts thereof shall be maintained in proper operating condition in accordance with the original design and in a safe condition. Devices or safeguards which are required by this code shall be maintained in compliance with the code under which they were installed.

102.4 Additions, alterations or repairs. Additions or alterations to installations shall conform to that required for new installations without requiring the existing installation to comply with all of the requirements of this code. Additions or alterations shall not cause an existing installation to become unsafe, hazardous or overloaded.

Repairs to existing installations shall be permitted in the same manner and arrangement as in the existing system and shall not be hazardous.

102.5 through 102.7 Reserved.

102.8 Referenced standards. The standards referenced in this code shall be those that are listed in Chapter 7 and such standards shall be considered part of the requirements of this code to the prescribed extent of each such reference. Where differences occur between provisions of this code and the referenced standards, the provisions of this code shall apply.

102.9 Other laws and regulations. The provisions of this code shall not be deemed to nullify any provisions of local, state or federal laws and regulations.

SECTION 103
OFFICE OF CODE ENFORCEMENT

103.1 General. A city, town, village or county that is responsible for administration and enforcement of this code shall designate a code enforcement official in accordance with the applicable provisions of local law.

A state agency that is responsible for administration and enforcement of this code shall be in compliance with the applicable provisions of state agency regulations.

SECTION 104
CODE ENFORCEMENT OFFICIAL

104.1 General. A city, town, village or county that is responsible for administration and enforcement of this code shall establish its local program in accordance with the applicable provisions of local law.

A state agency that is responsible for administration and enforcement of this code shall be in compliance with the applicable provisions of state agency regulations.

104.2 and 104.3 Reserved.

104.4 Inspections. A city, town, village or county that is responsible for administration and enforcement of this code shall provide for inspections in accordance with the applicable provisions of local law.

A state agency that is responsible for administration and enforcement of this code shall be in compliance with the applicable provisions of state agency regulations.

104.5 through 104.7 Reserved.

104.8 Department records. A city, town, village or county that is responsible for administration and enforcement of this code shall establish and maintain records in accordance with the applicable provisions of local law.

A state agency that is responsible for administration and enforcement of this code shall keep records in compliance with the applicable provisions of state agency regulations.

SECTION 105
APPROVAL

105.1 Reserved.

105.2 Alternative materials, methods and equipment. The provisions of this code are not intended to prevent the installation of any material or to prohibit any method of construction not specifically prescribed by this code, provided that any such alternative has been approved. An alternative material or method of construction shall be approved where the State Fire Prevention and Building Code Council finds that the proposed design is satisfactory and complies with the intent of the provisions of this code, and that the material, method or work offered is, for the purpose intended, at least the equivalent of that prescribed in this code in quality, strength, effectiveness, fire resistance, durability and safety.

105.3 Testing. Reserved.

105.3.1 Test methods. Test methods shall be as specified in this code or by other recognized test standards.

105.4 Material and equipment reuse. Material, equipment and devices shall not be reused unless they meet the requirements of this code for new materials.

SECTION 106
PERMITS

106.1 When required. A city, town, village or county that is responsible for administration and enforcement of this code shall determine local permit requirements in accordance with the applicable provisions of local law.

A state agency that is responsible for administration and enforcement of this code shall be in compliance with the applicable provisions of state agency regulations.

106.2 through 106.4 Reserved.

106.5 Fees. A city, town, village or county that is responsible for administration and enforcement of this code shall establish fees in accordance with the applicable provisions of local law.

A state agency that is responsible for administration and enforcement of this code shall be in compliance with the applicable provisions of state agency regulations.

SECTION 107
INSPECTIONS

107.1 Required inspections. A city, town, village or county that is responsible for administration and enforcement of this code shall determine necessary inspections in accordance that the applicable provisions of local law.

A state agency that is responsible for administration and enforcement of this code shall be in compliance with the applicable provisions of state agency regulations.

SECTION 108
VIOLATIONS

108.1 Violations. Violations of this code shall be dealt with in a manner appropriate to the applicable provisions of city, town, village or county and shall be in accordance with the applicable provisions of local law.

Violations of this code on state property shall be dealt with in a manner appropriate to a state agency.

108.2 through 108.6 Reserved.

108.7 Unsafe installations. An installation that is unsafe, constitutes a fire or health hazard, or is otherwise dangerous to human life, as regulated by this code, by reason of inadequate maintenance, dilapidation, fire hazard, disaster, damage or abandonment shall be abated by repair, rehabilitation, demolition or removal.

SECTION 109
VARIANCE PROCEDURES

109.1 Application for variance or appeal. Variance or appeal for any part of this code shall be in accordance with the provisions of 19 NYCRR titled, "Variance Procedures," which is administered by the Secretary of State. No town, village, city or county, nor any state agency charged with the administration and enforcement of this code, may waive, modify or otherwise alter this code.

CHAPTER 2
DEFINITIONS

SECTION 201
GENERAL

201.1 Scope. Unless otherwise expressly stated, the following words and terms shall, for the purposes of this code and standard, have the meanings indicated in this chapter.

201.2 Interchangeability. Words used in the present tense include the future; words in the masculine gender include the feminine and neuter; the singular number includes the plural and the plural, the singular.

201.3 Terms defined in other codes. Where terms are not defined in this code and are defined in the *Building Code of New York State*, *Fire Code of New York State*, *Mechanical Code of New York State* or *Plumbing Code of New York State*, such terms shall have meanings ascribed to them as in those codes.

201.4 Terms not defined. Where terms are not defined through the methods authorized by this section, such terms shall have ordinarily accepted meanings such as the context implies.

SECTION 202
GENERAL DEFINITIONS

ACCESS (TO). That which enables a device, appliance or equipment to be reached by ready access or by a means that first requires the removal or movement of a panel, door or similar obstruction (see also "Ready access").

AIR CONDITIONING. The treatment of air so as to control simultaneously the temperature, humidity, cleanness and distribution of the air to meet the requirements of a conditioned space.

AIR CONDITIONER, GAS-FIRED. A gas-burning, automatically operated appliance for supplying cooled and/or dehumidified air or chilled liquid.

AIR, EXHAUST. Air being removed from any space or piece of equipment and conveyed directly to the atmosphere by means of openings or ducts.

AIR-HANDLING UNIT. A blower or fan used for the purpose of distributing supply air to a room, space or area.

AIR, MAKEUP. Air that is provided to replace air being exhausted.

ALTERATION. A change in a system that involves an extension, addition or change to the arrangement, type or purpose of the original installation.

ANODELESS RISER. A transition assembly in which plastic piping is installed and terminated above ground outside of a building.

APPLIANCE (EQUIPMENT). Any apparatus or equipment that utilizes gas as a fuel or raw material to produce light, heat, power, refrigeration, or air conditioning.

APPLIANCE, AUTOMATICALLY CONTROLLED. Appliances equipped with an automatic burner ignition and safety shutoff device and other automatic devices which accomplish complete turn-on and shutoff of the gas to the main burner or burners, and graduate the gas supply to the burner or burners, but do not affect complete shutoff of the gas.

APPLIANCE, FAN-ASSISTED COMBUSTION. An appliance equipped with an integral mechanical means to either draw or force products of combustion through the combustion chamber or heat exchanger.

APPLIANCE, FUEL-FIRED. An appliance that burns solid, liquid and/or gaseous fuel, including but not limited to wood stoves, household cooking ranges, furnaces, boilers, water heaters, clothes dryers and gas-fired refrigerators.

APPLIANCE, GAS (EQUIPMENT). Any apparatus or equipment that uses gas as a fuel or raw material to produce light, heat, power, refrigeration or air conditioning.

APPLIANCE TYPE.

Low-heat appliance (residential appliance). Any appliance in which the products of combustion at the point of entrance to the flue under normal operating conditions have a temperature of 1,000°F (538°C) or less.

Medium-heat appliance. Any appliance in which the products of combustion at the point of entrance to the flue under normal operating conditions have a temperature of more than 1,000°F (538°C), but not greater than 2,000°F (1093°C).

APPLIANCE, UNVENTED. An appliance designed or installed in such a manner that the products of combustion are not conveyed by a vent or chimney directly to the outside atmosphere.

APPLIANCE, VENTED. An appliance designed and installed in such a manner that all of the products of combustion are conveyed directly from the appliance to the outside atmosphere through an approved chimney or vent system.

APPROVED. Acceptable to the code enforcement official.

APPROVED AGENCY. An established and recognized agency that is regularly engaged in conducting tests or furnishing inspection services.

ATMOSPHERIC PRESSURE. The pressure of the weight of air and water vapor on the surface of the earth, approximately 14.7 pounds per square inch (psi) (101 kPa absolute) at sea level.

AUTHORITY HAVING JURISDICTION. The local government, county government or state agency responsible for the administration and enforcement of an applicable regulation or law.

AUTOMATIC IGNITION. Ignition of gas at the burner(s) when the gas controlling device is turned on, including reignition if the flames on the burner(s) have been extinguished by means other than by the closing of the gas controlling device.

BAFFLE. An object placed in an appliance to change the direction of or retard the flow of air, air-gas mixtures, or flue gases.

BAROMETRIC DRAFT REGULATOR. A balanced damper device attached to a chimney, vent connector, breeching, or flue gas manifold to protect combustion equipment by controlling chimney draft. A double-acting barometric draft regulator is one whose balancing damper is free to move in either direction to protect combustion equipment from both excessive draft and backdraft.

BOILER, LOW-PRESSURE. A self-contained appliance for supplying steam or hot water.

Hot water heating boiler. A boiler in which no steam is generated, from which hot water is circulated for heating purposes and then returned to the boiler, and that operates at water pressures not exceeding 160 psig (1100 kPa gauge) and at water temperatures not exceeding 250°F (121°C) at or near the boiler outlet.

Hot water supply boiler. A boiler, completely filled with water, which furnishes hot water to be used externally to itself, and that operates at water pressures not exceeding 160 psig (1100 kPa gauge) and at water temperatures not exceeding 250°F (121°C) at or near the boiler outlet.

Steam heating boiler. A boiler in which steam is generated and that operates at a steam pressure not exceeding 15 psig (100 kPa gauge).

BRAZING. A metal-joining process wherein coalescence is produced by the use of a nonferrous filler metal having a melting point above 1,000°F (538°C), but lower than that of the base metal being joined. The filler material is distributed between the closely fitted surfaces of the joint by capillary action.

BROILER. A general term including salamanders, barbecues, and other appliances cooking primarily by radiated heat, excepting toasters.

BTU. Abbreviation for British thermal unit, which is the quantity of heat required to raise the temperature of 1 pound (454 g) of water 1°F (1.8°C) (1 Btu = 1055 J).

BURNER. A device for the final conveyance of the gas, or a mixture of gas and air, to the combustion zone.

Induced-draft. A burner that depends on draft induced by a fan that is an integral part of the appliance and is located downstream from the burner.

Power. A burner in which gas, air or both are supplied at pressures exceeding, for gas, the line pressure, and for air, atmospheric pressure, with this added pressure being applied at the burner.

CHIMNEY. A primarily vertical structure containing one or more flues, for the purpose of carrying gaseous products of combustion and air from an appliance to the outside atmosphere.

Factory-built chimney. A listed and labeled chimney composed of factory-made components, assembled in the field in accordance with manufacturer's instructions and the conditions of the listing.

Masonry chimney. A field-constructed chimney composed of solid masonry units, bricks, stones or concrete.

Metal chimney. A field-constructed chimney of metal.

CLEARANCE. The minimum distance through air measured between the heat-producing surface of the mechanical appliance, device or equipment and the surface of the combustible material or assembly.

CLOTHES DRYER. An appliance used to dry wet laundry by means of heated air. Dryer classifications are as follows:

Type 1. Factory-built package, multiple production. Primarily used in family living environment. Usually the smallest unit physically and in function output.

Type 2. Factory-built package, multiple production. Used in business with direct intercourse of the function with the public. Not designed for use in individual family living environment.

CODE. These regulations, subsequent amendments thereto, or any emergency rule or regulation that the administrative authority having jurisdiction has lawfully adopted.

CODE ENFORCEMENT OFFICIAL. The officer or other designated authority charged with the administration and enforcement of this code, or a duly authorized representative.

COMBUSTION. In the context of this code, refers to the rapid oxidation of fuel accompanied by the production of heat or heat and light.

COMBUSTION AIR. Air necessary for complete combustion of a fuel, including theoretical air and excess air.

COMBUSTION CHAMBER. The portion of an appliance within which combustion occurs.

COMBUSTION PRODUCTS. Constituents resulting from the combustion of a fuel with the oxygen of the air, including the inert gases, but excluding excess air.

CONCEALED LOCATION. A location that cannot be accessed without damaging permanent parts of the building structure or finish surface. Spaces above, below or behind readily removable panels or doors shall not be considered as concealed.

CONCEALED PIPING. Piping that is located in a concealed location (see "Concealed location").

CONDENSATE. The liquid that condenses from a gas (including flue gas) caused by a reduction in temperature or increase in pressure.

CONFINED SPACES. A space having a volume less than 50 cubic feet per 1,000 British thermal units per hour (Btu/h) (4.8 m³/kW) of the aggregate input rating of all appliances installed in that space.

CONNECTOR. The pipe that connects an approved appliance to a chimney, flue or vent.

CONSTRUCTION DOCUMENTS. All of the written, graphic and pictorial documents prepared or assembled for describing the design, location and physical characteristics of the elements of the project necessary for obtaining a mechanical permit.

CONTROL. A manual or automatic device designed to regulate the gas, air, water or electrical supply to, or operation of, a mechanical system.

CONVERSION BURNER. A unit consisting of a burner and its controls for installation in an appliance originally utilizing another fuel.

COUNTER APPLIANCES. Appliances such as coffee brewers and coffee urns and any appurtenant water-heating equipment, food and dish warmers, hot plates, griddles, waffle bakers and other appliances designed for installation on or in a counter.

CUBIC FOOT. The amount of gas that occupies 1 cubic foot (0.02832 m³) when at a temperature of 60°F (16°C), saturated with water vapor and under a pressure equivalent to that of 30 inches of mercury (101 kPa).

DAMPER. A manually or automatically controlled device to regulate draft or the rate of flow of air or combustion gases.

DECORATIVE APPLIANCE, VENTED. A vented appliance wherein the primary function lies in the aesthetic effect of the flames.

DECORATIVE APPLIANCES FOR INSTALLATION IN VENTED FIREPLACES. A vented appliance designed for installation within the fire chamber of a vented fireplace, wherein the primary function lies in the aesthetic effect of the flames.

DEMAND. The maximum amount of gas input required per unit of time, usually expressed in cubic feet per hour, or Btu/h (1 Btu/h = 0.2931 W).

DILUTION AIR. Air that is introduced into a draft hood and is mixed with the flue gases.

DIRECT-FIRED INDUSTRIAL AIR HEATER. A heater in which all of the products of combustion generated by the burners are released into the airstream being heated; whose purpose is to offset the building heat loss by heating incoming outside air, inside air or a combination of both.

DIRECT-FIRED MAKEUP AIR HEATER. A heater in which all of the products of combustion generated by the burners are released into the outdoor airstream being heated.

DIRECT-VENT APPLIANCES. Appliances that are constructed and installed so that all air for combustion is derived directly from the outside atmosphere and all flue gases are discharged directly to the outside atmosphere.

DRAFT. The pressure difference existing between the equipment or any component part and the atmosphere, that causes a continuous flow of air and products of combustion through the gas passages of the appliance to the atmosphere.

    Mechanical or induced draft. The pressure difference created by the action of a fan, blower or ejector, that is located between the appliance and the chimney or vent termination.

Natural draft. The pressure difference created by a vent or chimney because of its height, and the temperature difference between the flue gases and the atmosphere.

DRAFT HOOD. A nonadjustable device built into an appliance, or made as part of the vent connector from an appliance, that is designed to (1) provide for ready escape of the flue gases from the appliance in the event of no draft, backdraft, or stoppage beyond the draft hood, (2) prevent a backdraft from entering the appliance, and (3) neutralize the effect of stack action of the chimney or gas vent upon operation of the appliance.

DRAFT REGULATOR. A device that functions to maintain a desired draft in the appliance by automatically reducing the draft to the desired value.

DRIP. The container placed at a low point in a system of piping to collect condensate and from which the condensate is removable.

DRY GAS. A gas having a moisture and hydrocarbon dew point below any normal temperature to which the gas piping is exposed.

DUCT FURNACE. A warm-air furnace normally installed in an air distribution duct to supply warm air for heating. This definition shall apply only to a warm-air heating appliance that depends for air circulation on a blower not furnished as part of the furnace.

DUCT SYSTEM. A continuous passageway for the transmission of air that, in addition to ducts, includes duct fittings, dampers, plenums, fans and accessory air-handling equipment.

EQUIPMENT. See "Appliance."

F RATING. The time period that the through-penetration firestop system limits the spread of fire through the penetration when tested in accordance with ASTM E 814.

FIRE-RESISTANCE RATING. The period of time a building element, component or assembly maintains the ability to confine a fire, continues to perform a given structural function, or both as determined by the tests, or the methods based on tests, prescribed in Section 703 of the *Building Code of New York State.*

FIREPLACE. A fire chamber and hearth constructed of noncombustible material for use with solid fuels and provided with a chimney.

Masonry fireplace. A hearth and fire chamber of solid masonry units such as bricks, stones, listed masonry units, or reinforced concrete, provided with a suitable chimney.

Factory-built fireplace. A fireplace composed of listed factory-built components assembled in accordance with the terms of listing to form the completed fireplace.

FIRING VALVE. A valve of the plug and barrel type designed for use with gas, and equipped with a lever handle for manual operation and a dial to indicate the percentage of opening.

FLAME SAFEGUARD. A device that will automatically shut off the fuel supply to a main burner or group of burners when the means of ignition of such burners become inoperative, and when flame failure occurs on the burner or group of burners.

FLOOR FURNACE. A completely self-contained furnace suspended from the floor of the space being heated, taking air for combustion from outside such space and with means for observing flames and lighting the appliance from such space.

Gravity type. A floor furnace depending primarily upon circulation of air by gravity. This classification shall also include floor furnaces equipped with booster type fans which do not materially restrict free circulation of air by gravity flow when such fans are not in operation.

Fan type. A floor furnace equipped with a fan which provides the primary means for circulating air.

FLUE, APPLIANCE. The passage(s) within an appliance through which combustion products pass from the combustion chamber of the appliance to the draft hood inlet opening on an appliance equipped with a draft hood or to the outlet of the appliance on an appliance not equipped with a draft hood.

FLUE COLLAR. That portion of an appliance designed for the attachment of a draft hood, vent connector, or venting system.

FLUE GASES. Products of combustion plus excess air in appliance flues or heat exchangers.

FLUE LINER (LINING). A system or material used to form the inside surface of a flue in a chimney or vent, for the purpose of protecting the surrounding structure from the effects of combustion products and for conveying combustion products without leakage to the atmosphere.

FUEL GAS. A natural, manufactured, liquefied petroleum or a mixture of these.

FUEL GAS UTILIZATION EQUIPMENT. See "Appliance."

FURNACE. A completely self-contained heating unit that is designed to supply heated air to spaces remote from or adjacent to the appliance location.

FURNACE, CENTRAL. A self-contained appliance for heating air by transfer of heat of combustion through metal to the air, and designed to supply heated air through ducts to spaces remote from or adjacent to the appliance location.

Gravity type. A central furnace depending primarily on circulation of air by gravity.

Gravity furnace with booster fan. A furnace equipped with a booster fan that does not materially restrict free circulation of air by gravity flow when the fan is not in operation.

Forced-air furnace with cooling unit. A single-package unit, consisting of a gas-fired forced-air furnace of one of the types listed below combined with an electrically or fuel gas-powered summer air-conditioning system, contained in a common casing.

Forced-air type. A central furnace equipped with a fan or blower which provides the primary means for circulation of air.

Horizontal forced-air type. A furnace with airflow through the appliance essentially in a horizontal path.

Downflow furnace. A furnace designed with airflow discharge vertically downward at or near the bottom of the furnace.

Upflow furnace. A furnace designed with airflow discharge vertically upward at or near the top of the furnace. This classification includes "highboy" furnaces with the blower mounted below the heating element and "lowboy" furnaces with the blower mounted beside the heating element.

Multiple-position furnace. A furnace designed so that it can be installed with the airflow discharge in the upflow, horizontal or downflow direction.

FURNACE, ENCLOSED. A specific heating, or heating and ventilating, furnace incorporating an integral total enclosure and using only outside air for combustion.

GAS CONVENIENCE OUTLET. A permanently mounted, manually operated device that provides the means for connecting an appliance to, and disconnecting an appliance from, the supply piping. The device includes an integral, manually operated valve with a nondisplaceable valve member and is designed so that disconnection of an appliance only occurs when the manually operated valve is in the closed position.

GAS PIPING. An installation of pipe, valves or fittings installed on a premises or in a building and utilized to convey fuel gas.

GAS UTILIZATION EQUIPMENT. An appliance that utilizes gas as a fuel or raw material or both.

HAZARDOUS LOCATION. Any location considered to be a fire hazard for flammable vapors, dust, combustible fibers or other highly combustible substances. The location is not necessarily categorized in the building code as a high-hazard use group classification.

HOUSE PIPING. See "Piping system."

IGNITION PILOT. A pilot that operates during the lighting cycle and discontinues during main burner operation.

IGNITION SOURCE. A flame, spark or hot surface capable of igniting flammable vapors or fumes. Such sources include appliance burners, burner ignitors, and electrical switching devices.

INCINERATOR. An appliance used to reduce combustible refuse material to ashes and which is manufactured, sold and installed as a complete unit.

INFRARED RADIANT HEATER. A heater that directs a substantial amount of its energy output in the form of infrared radiant energy into the area to be heated. Such heaters are of either the vented or unvented type.

JOINT, FLANGED. A joint made by bolting together a pair of flanged ends.

JOINT, FLARED. A metal-to-metal compression joint in which a conical spread is made on the end of a tube that is compressed by a flare nut against a mating flare.

JOINT, MECHANICAL. A general form of gas-tight joints obtained by the joining of metal parts through a positive-holding mechanical construction, such as flanged joint, threaded joint, flared joint, or compression joint.

JOINT, PLASTIC ADHESIVE. A joint made in thermoset plastic piping by the use of an adhesive substance which forms a continuous bond between the mating surfaces without dissolving either one of them.

JOINT, PLASTIC HEAT FUSION. A joint made in thermoplastic piping by heating the parts sufficiently to permit fusion of the materials when the parts are pressed together.

JOINT, WELDED. A gas-tight joint obtained by the joining of metal parts in molten state.

LABELED. Devices, equipment, appliances or materials to which have been affixed a label, seal, symbol or other identifying mark of a nationally recognized testing laboratory, inspection agency or other organization concerned with product evaluation that maintains periodic inspection of the production of the above-labeled items and by whose label the manufacturer attests to compliance with applicable nationally recognized standards.

LIMIT CONTROL. A device responsive to changes in pressure, temperature or level for turning on, shutting off or throttling the gas supply to an appliance.

LIQUEFIED PETROLEUM GAS or LPG (LP-GAS). Liquefied petroleum gas composed predominately of propane, propylene, butanes or butylenes, or mixtures thereof that is gaseous under normal atmospheric conditions, but is capable of being liquefied under moderate pressure at normal temperatures.

LISTED. Equipment, appliances or materials included in a list published by a nationally recognized testing laboratory, inspection agency or other organization concerned with product evaluation that maintains periodic inspection of production of listed equipment, appliances or materials, and whose listing states either that the equipment, appliance or material meets nationally recognized standards or has been tested and found suitable for use in a specified manner. The means for identifying listed equipment, appliances or materials may vary for each testing laboratory, inspection agency, or other organization concerned with product evaluation, some of which do not recognize equipment, appliances or materials as listed unless it is also labeled. The authority having jurisdiction shall utilize the system employed by the listing organization to identify a listed product.

LIVING SPACE. Space within a dwelling unit utilized for living, sleeping, eating, cooking, bathing, washing and sanitation purposes.

LOG LIGHTER. A manually operated solid fuel ignition appliance for installation in a vented solid fuel-burning fireplace.

LUBRICATED PLUG-TYPE VALVE. A valve of the plug and barrel type provided with means for maintaining a lubricant between the bearing surfaces.

MAIN BURNER. A device or group of devices essentially forming an integral unit for the final conveyance of gas or a mixture of gas and air to the combustion zone, and on which combustion takes place to accomplish the function for which the appliance is designed.

MECHANICAL EXHAUST SYSTEM. Equipment installed in and made a part of the vent, which will provide a positive induced draft.

MEMBRANE PENETRATION. An opening made through one side (wall, floor or ceiling membrane) of an assembly.

MEMBRANE-PENETRATION FIRESTOP. A material device or construction installed to resist for a prescribed time period, the passage of flame and heat through openings in a protective membrane in order to accommodate cables, cable trays, conduit, tubing, pipes or similar items.

METER. The instrument installed to measure the volume of gas delivered through it.

MODULATING. Modulating or throttling is the action of a control from its maximum to minimum position in either predetermined steps or increments of movement as caused by its actuating medium.

OCCUPANCY. The purpose for which a building, or portion thereof, is utilized or occupied.

OFFSET (VENT). A combination of approved bends that makes two changes in direction bringing one section of the vent out of line but into a line parallel with the other section.

ORIFICE. The opening in a cap, spud or other device whereby the flow of gas is limited and through which the gas is discharged to the burner.

OUTLET. A threaded connection or bolted flange in a pipe system to which a gas-burning appliance is attached.

OXYGEN DEPLETION SAFETY SHUTOFF SYSTEM (ODS). A system designed to act to shut off the gas supply to the main and pilot burners if the oxygen in the surrounding atmosphere is reduced below a predetermined level.

PENETRATION FIRESTOP. A through-penetration firestop or a membrane-penetration firestop.

PILOT. A small flame that is utilized to ignite the gas at the main burner or burners.

PIPING. Where used in this code, "piping" refers to either pipe or tubing, or both.

    Pipe. A rigid conduit of iron, steel, copper, brass or plastic.

    Tubing. Semirigid conduit of copper, aluminum, plastic or steel.

PIPING SYSTEM. All fuel piping, valves, and fittings from the outlet of the point of delivery to the connections with the gas utilization equipment.

PLASTIC, THERMOPLASTIC. A plastic that is capable of being repeatedly softened by increase of temperature and hardened by decrease of temperature.

PLENUM. Air compartment or chamber to which one or more ducts are connected and which forms part of an air distribution system.

POINT OF DELIVERY. The point of delivery is the outlet of the service meter assembly, or the outlet of the service regulator or service shutoff valve where a meter is not provided.

For undiluted liquefied petroleum gas systems, the point of delivery shall be considered the outlet of the first-stage pressure regulator that provides utilization pressure, exclusive of line gas regulators, in the system.

PRESSURE DROP. The loss in pressure due to friction or obstruction in pipes, valves, fittings, regulators, and burners.

PRESSURE TEST. An operation performed to verify the gas-tight integrity of gas piping following its installation or modification.

PURGE. To free a gas conduit of air or gas, or a mixture of gas and air.

QUICK-DISCONNECT DEVICE. A hand-operated device that provides a means for connecting and disconnecting an appliance or an appliance connector to a gas supply and that is equipped with an automatic means to shut off the gas supply when the device is disconnected.

READY ACCESS (TO). That which enables a device, appliance or equipment to be directly reached, without requiring the removal or movement of any panel, door or similar obstruction (see "Access").

REGISTERED DESIGN PROFESSIONAL. An individual who is a registered architect (RA) in accordance with Article 147 of the New York State Education Law or a licensed professional engineer (PE) in accordance with Article 145 of the New York State Education Law.

REGULATOR. A device for controlling and maintaining a uniform supply pressure, either pounds-to-inches water column (MP regulator) or inches-to-inches water column (appliance regulator).

REGULATOR, GAS APPLIANCE. A pressure regulator for controlling pressure to the manifold of equipment. Types of appliance regulators are as follows:

Adjustable.
1. Spring type, limited adjustment. A regulator in which the regulating force acting upon the diaphragm is derived principally from a spring, the loading of which is adjustable over a range of not more than 15 percent of the outlet pressure at the midpoint of the adjustment range.
2. Spring type, standard adjustment. A regulator in which the regulating force acting upon the diaphragm is derived principally from a spring, the loading of which is adjustable. The adjustment means shall be concealed.

Multistage. A regulator for use with a single gas whose adjustment means is capable of being positioned manually or automatically to two or more predetermined outlet pressure settings. Each of these settings shall be adjustable or nonadjustable. The regulator may modulate outlet pressures automatically between its maximum and minimum predetermined outlet pressure settings.

Nonadjustable.
1. Spring type, nonadjustable. A regulator in which the regulating force acting upon the diaphragm is derived principally from a spring, the loading of which is not field adjustable.
2. Weight type. A regulator in which the regulating force acting upon the diaphragm is derived from a weight or combination of weights.

REGULATOR, LINE GAS PRESSURE. A device placed in a gas line between the service pressure regulator and the equipment for controlling, maintaining or reducing the pressure in that portion of the piping system downstream of the device.

REGULATOR, MEDIUM-PRESSURE. A medium-pressure (MP) regulator reduces the gas piping pressure to the appliance regulator or to the appliance utilization pressure.

REGULATOR, PRESSURE. A device placed in a gas line for reducing, controlling, and maintaining the pressure in that portion of the piping system downstream of the device.

REGULATOR, SERVICE PRESSURE. A device installed by the serving gas supplier to reduce and limit the service line pressure to delivery pressure.

RELIEF OPENING. The opening provided in a draft hood to permit the ready escape to the atmosphere of the flue products from the draft hood in the event of no draft, back draft, or stoppage beyond the draft hood, and to permit air into the draft hood in the event of a strong chimney updraft.

RELIEF VALVE (DEVICE). A safety valve designed to forestall the development of a dangerous condition by relieving either pressure, temperature, or vacuum in the hot water supply system.

RELIEF VALVE, PRESSURE. An automatic valve that opens and closes a relief vent, depending on whether the pressure is above or below a predetermined value.

RELIEF VALVE, TEMPERATURE
Reseating or self-closing type. An automatic valve that opens and closes a relief vent, depending on whether the temperature is above or below a predetermined value.

Manual reset type. A valve that automatically opens a relief vent at a predetermined temperature and that must be manually returned to the closed position.

RELIEF VALVE, VACUUM. A valve that automatically opens and closes a vent for relieving a vacuum within the hot water supply system, depending on whether the vacuum is above or below a predetermined value.

RISER, GAS. A vertical pipe supplying fuel gas.

ROOM HEATER, UNVENTED. See "Unvented room heater."

ROOM HEATER, VENTED. A free-standing heating unit used for direct heating of the space in and adjacent to that in which the unit is located (see also "Vented room heater").

ROOM LARGE IN COMPARISON WITH SIZE OF EQUIPMENT. Rooms having a volume equal to at least 12 times the total volume of a furnace or air-conditioning appliance and at least 16 times the total volume of a boiler. Total volume of the appliance is determined from exterior dimensions and is to include fan compartments and burner vestibules, when used. When the actual ceiling height of a room is greater than 8 feet (2438 mm), the volume of the room is figured on the basis of a ceiling height of 8 feet (2438 mm).

SAFETY SHUTOFF DEVICE. See "Flame safeguard."

SHAFT. An enclosed space extending through one or more stories of a building, connecting vertical openings in successive floors, or floors and the roof.

SPECIFIC GRAVITY. As applied to gas, specific gravity is the ratio of the weight of a given volume to that of the same volume of air, both measured under the same condition.

T RATING. The time period that the penetration firestop system, including the penetrating item, limits the maximum temperature rise to 325°F (163°C) above its initial temperature through the penetration on the nonfire side when tested in accordance with ASTM E 814.

THERMOSTAT.

> **Electric switch type.** A device that senses changes in temperature and controls electrically, by means of separate components, the flow of gas to the burner(s) to maintain selected temperatures.

> **Integral gas valve type.** An automatic device, actuated by temperature changes, designed to control the gas supply to the burner(s) in order to maintain temperatures between predetermined limits, and in which the thermal actuating element is an integral part of the device.
> 1. Graduating thermostat. A thermostat in which the motion of the valve is approximately in direct proportion to the effective motion of the thermal element induced by temperature change.
> 2. Snap-acting thermostat. A thermostat in which the thermostatic valve travels instantly from the closed to the open position, and vice versa.

THROUGH PENETRATION. An opening that passes through an entire assembly.

THROUGH-PENETRATION FIRESTOP SYSTEM. An assemblage of specific materials or products that are designed, tested and fire-resistance rated to resist for a prescribed period of time the spread of fire through penetrations. The F and T rating criteria for penetration firestop systems shall be in accordance with ASTM E 814. See definition of "F rating" and "T rating."

TRANSITION FITTINGS, PLASTIC TO STEEL. An adapter for joining plastic pipe to steel pipe. The purpose of this fitting is to provide a permanent, pressure-tight connection between two materials which cannot be joined directly one to another.

UNCONFINED SPACE. A space having a volume not less than 50 cubic feet per 1,000 Btu/h (4.8 m^3/kW) of the aggregate input rating of all appliances installed in that space. Rooms communicating directly with the space in which the appliances are installed, through openings not furnished with doors, are considered a part of the unconfined space.

UNIT HEATER.

> **High-static pressure type.** A self-contained, automatically controlled, vented appliance having integral means for circulation of air against 0.2 inch (15 mm H_2O) or greater static pressure. Such appliance is equipped with provisions for attaching an outlet air duct and, where the appliance is for indoor installation remote from the space to be heated, is also equipped with provisions for attaching an inlet air duct.

> **Low-static pressure type.** A self-contained, automatically controlled, vented appliance, intended for installation in the space to be heated without the use of ducts, having integral means for circulation of air. Such units are allowed to be equipped with louvers or face extensions made in accordance with the manufacturer's specifications.

UNLISTED BOILER. A boiler not listed by a nationally recognized testing agency.

UNUSUALLY TIGHT CONSTRUCTION. Construction meeting the following requirements:
1. Walls and ceilings exposed to the outside atmosphere having a continuous water vapor retarder with a rating of 1 perm (5.72 x 10^{-8} g/Pa • s • m^2) or less with openings gasketed or sealed; and
2. Storm windows or weatherstripping on openable windows and doors;
3. Caulking or sealants applied to areas, such as joints around window and door frames, between sole plates and floors, between wall-ceiling joints, between wall panels, at penetrations for plumbing, electrical and gas lines, and at other openings.

UNVENTED ROOM HEATER. An unvented heating appliance designed for stationary installation and utilized to provide comfort heating. Such appliances provide radiant heat or convection heat by gravity or fan circulation directly from the heater and do not utilize ducts. A wall-mounted unvented room heater would be of the type designed for insertion in or attachment to a wall or partition. A wall-mounted unvented room heater does not incorporate concealed venting arrangements in its construction and discharges all products of combustion through the front into the room being heated.

VALVE. A device used in piping to control the gas supply to any section of a system of piping or to an appliance.

Automatic. An automatic or semiautomatic device consisting essentially of a valve and operator that control the gas supply to the burner(s) during operation of an appliance. The operator shall be actuated by application of gas pressure on a flexible diaphragm or by electrical means or by mechanical means.

Automatic gas shutoff. A valve used in conjunction with an automatic gas shutoff device to shut off the gas supply to a water-heating system. It shall be constructed integrally with the gas shutoff device or shall be a separate assembly.

Equipment shutoff. A valve located in the piping system, used to isolate individual equipment for purposes such as service or replacement.

Individual main burner. A valve that controls the gas supply to an individual main burner.

Main burner control. A valve that controls the gas supply to the main burner manifold.

Manual main gas-control. A manually operated valve in the gas line for the purpose of completely turning on or shutting off the gas supply to the appliance, except to pilot or pilots that are provided with independent shutoff.

Manual reset. An automatic shutoff valve installed in the gas supply piping and set to shut off when unsafe conditions occur. The device remains closed until manually reopened.

Service shutoff. A valve, installed by the serving gas supplier between the service meter or source of supply and the customer piping system, to shut off the entire piping system.

VENT. A pipe or other conduit composed of factory-made components, containing a passageway for conveying combustion products and air to the atmosphere, listed and labeled for use with a specific type or class of appliance.

Special gas vent. A vent listed and labeled for use with listed Category II, III and IV appliances.

Type B vent. A vent listed and labeled for use with appliances with draft hoods and other Category I appliances that are listed for use with Type B vents.

Type BW vent. A vent listed and labeled for use with wall furnaces.

Type L vent. A vent listed and labeled for use with appliances that are listed for use with Type L or Type B vents.

VENT CONNECTOR. See "Connector."

VENT GASES. Products of combustion from appliances plus excess air plus dilution air in the vent connector, gas vent or chimney above the draft hood or draft regulator.

VENTED APPLIANCE CATEGORIES. Appliances that are categorized for the purpose of vent selection are classified into the following four categories:

Category I. An appliance that operates with a nonpositive vent static pressure and with a vent gas temperature that avoids excessive condensate production in the vent.

Category II. An appliance that operates with a nonpositive vent static pressure and with a vent gas temperature that is capable of causing excessive condensate production in the vent.

Category III. An appliance that operates with a positive vent static pressure and with a vent gas temperature that avoids excessive condensate production in the vent.

Category IV. An appliance that operates with a positive vent static pressure and with a vent gas temperature that is capable of causing excessive condensate production in the vent.

VENTED ROOM HEATER. A vented self-contained, free-standing, nonrecessed appliance for furnishing warm air to the space in which it is installed, directly from the heater without duct connections.

VENTED WALL FURNACE. A self-contained vented appliance complete with grilles or equivalent, designed for incorporation in or permanent attachment to the structure of a building, mobile home or travel trailer, and furnishing heated air circulated by gravity or by a fan directly into the space to be heated through openings in the casing. This definition shall exclude floor furnaces, unit heaters and central furnaces as herein defined.

VENTING SYSTEM. A continuous open passageway from the flue collar or draft hood of an appliance to the outside atmosphere for the purpose of removing flue or vent gases. A venting system is usually composed of a vent or a chimney and vent connector, if used, assembled to form the open passageway.

Mechanical draft venting system. A venting system designed to remove flue or vent gases by mechanical means, that consists of an induced draft portion under nonpositive static pressure or a forced draft portion under positive static pressure.

> **a. Forced-draft venting system.** A portion of a venting system using a fan or other mechanical means to cause the removal of flue or vent gases under positive static vent pressure.
>
> **b. Induced draft venting system.** A portion of a venting system using a fan or other mechanical means to cause the removal of flue or vent gases under nonpositive static vent pressure.
>
> **c. Power venting system.** A portion of a venting system using a fan or other mechanical means to cause the removal of flue or vent gases under positive static vent pressure.

Natural draft venting system. A venting system designed to remove flue or vent gases under nonpositive static vent pressure entirely by natural draft.

WALL HEATER, UNVENTED-TYPE. A room heater of the type designed for insertion in or attachment to a wall or partition. Such heater does not incorporate concealed venting arrangements in its construction and discharges all products of combustion through the front into the room being heated.

WATER HEATER. Any heating appliance or equipment that heats potable water and supplies such water to the potable hot water distribution system.

CHAPTER 3
GENERAL REGULATIONS

SECTION 301
GENERAL

301.1 Scope. This chapter shall govern the approval and installation of all equipment and appliances that comprise parts of the installations regulated by this code in accordance with Section 101.2.

301.1.1 Other fuels. The requirements for combustion and dilution air for gas-fired appliances shall be governed by Section 304. The requirements for combustion and dilution air for appliances operating with fuels other than fuel gas shall be regulated by the *Mechanical Code of New York State*.

301.2 Energy utilization. Heating, ventilating and air-conditioning systems of all structures shall be designed and installed for efficient utilization of energy in accordance with the *Energy Conservation Construction Code of New York State*.

301.3 Listed and labeled. Appliances regulated by this code shall be listed and labeled.

301.4 Labeling. Labeling shall be in accordance with the procedures set forth in Sections 301.4.1 through 301.4.2.3.

301.4.1 Testing. An approved agency shall test a representative sample of the appliances being labeled to the relevant standard or standards. The approved agency shall maintain a record of all of the tests performed. The record shall provide sufficient detail to verify compliance with the test standard.

301.4.2 Inspection and identification. The approved agency shall periodically perform an inspection, which shall be in-plant if necessary, of the appliances to be labeled. The inspection shall verify that the labeled appliances are representative of the appliances tested.

301.4.2.1 Independent. The agency to be approved shall be objective and competent. To confirm its objectivity, the agency shall disclose all possible conflicts of interest.

301.4.2.2 Equipment. An approved agency shall have adequate equipment to perform all required tests. The equipment shall be periodically calibrated.

301.4.2.3 Personnel. An approved agency shall employ experienced personnel educated in conducting, supervising and evaluating tests.

301.5 Label information. A permanent factory-applied nameplate(s) shall be affixed to appliances on which shall appear in legible lettering, the manufacturer's name or trademark, the model number, serial number and, for listed appliances, the seal or mark of the testing agency. A label shall also include the hourly rating in Btu/h (W); the type of fuel approved for use with the appliance; and the minimum clearance requirements.

301.6 Plumbing connections. Potable water supply and building drainage system connections to appliances regulated by this code shall be in accordance with the *Plumbing Code of New York State*.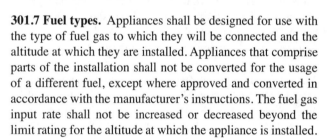

301.7 Fuel types. Appliances shall be designed for use with the type of fuel gas to which they will be connected and the altitude at which they are installed. Appliances that comprise parts of the installation shall not be converted for the usage of a different fuel, except where approved and converted in accordance with the manufacturer's instructions. The fuel gas input rate shall not be increased or decreased beyond the limit rating for the altitude at which the appliance is installed.

301.8 Vibration isolation. Where means for isolation of vibration of an appliance is installed, an approved means for support and restraint of that appliance shall be provided.

301.9 Repair. Defective material or parts shall be replaced or repaired in such a manner so as to preserve the original approval or listing.

301.10 Wind resistance. Appliances and supports that are exposed to wind shall be designed and installed to resist the wind pressures determined in accordance with the *Building Code of New York State*.

301.11 Flood hazard. For structures located in special flood hazard areas, the appliance, equipment and system installations regulated by this code shall comply with the flood-resistant construction requirements of the *Building Code of New York State*.

301.12 Seismic resistance. When earthquake loads are applicable in accordance with the *Building Code of New York State*, the supports shall be designed and installed for the seismic forces in accordance with *Section 1621* of the *Building Code of New York State*.

301.13 Ducts. All ducts required for the installation of systems regulated by this code shall be designed and installed in accordance with the *Mechanical Code of New York State*.

301.14 Rodentproofing. Buildings or structures and the walls enclosing habitable or occupiable rooms and spaces in which persons live, sleep or work, or in which feed, food or foodstuffs are stored, prepared, processed, served or sold, shall be constructed to protect against rodents in accordance with the *Building Code of New York State*.

301.15 Prohibited location. The appliances, equipment and systems regulated by this code shall not be located in an elevator shaft.

SECTION 302
STRUCTURAL SAFETY

302.1 Structural safety. The building shall not be weakened by the installation of any gas piping. In the process of installing or repairing any gas piping, the finished floors, walls, ceilings, tile work or any other part of the building or premises which are required to be changed or replaced shall be left in a safe structural condition in accordance with the requirements of Sections 302.2 through 302.4.

302.2 Penetrations of floor/ceiling assemblies and fire-resistance-rated assemblies. Penetrations of floor/ceiling assemblies and assemblies required to have a fire-resistance rating shall be protected in accordance with Sections 302.2.1 through 302.2.3.

302.2.1 Installation details. Where sleeves are used, they shall be fastened securely to the assembly penetrated. The space between the item contained in the sleeve and the sleeve itself and any space between the sleeve and the assembly penetrated shall be protected in accordance with this section. Insulation and coverings on or in the penetrating item shall not penetrate the assembly unless the specific material used has been tested as part of the assembly in accordance with this section.

302.2.2 Fire-resistance-rated walls. Penetrations into or through fire walls, fire barriers, smoke barrier walls, and fire partitions shall comply with this section.

302.2.2.1 Through penetrations. Through penetration of fire-resistance-rated walls shall comply with Section 302.2.2.1.1 or Section 302.2.2.1.2.

Exception: Where the penetrating items are steel, ferrous or copper pipes or steel conduits, the annular space between the penetrating item and the fire-resistance-rated wall shall be permitted to be protected as follows:

1. In concrete or masonry walls where the penetrating item is a maximum 6-inch (152 mm) nominal diameter and the opening is a maximum 144 square inches (92 900 mm^2), concrete, grout or mortar shall be permitted where installed the full thickness of the wall or the thickness required to maintain the fire-resistance rating; or

2. The material used to fill the annular space shall prevent the passage of flame and hot gases sufficient to ignite cotton waste where subjected to ASTM E 119 time temperature fire conditions under a minimum positive pressure differential of 0.01 inch (2.49 Pa) of water at the location of the penetration for the time period equivalent to the fire-resistance rating of the construction penetrated.

302.2.2.1.1 Fire-resistance-rated assemblies. Penetrations shall be installed as tested in an approved fire-resistance-rated assembly.

302.2.2.1.2 Through-penetration firestop system. Through penetrations shall be protected by an approved penetration firestop system installed as tested in accordance with ASTM E 814, with a minimum positive pressure differential of 0.01 inch (2.49 Pa) of water, and shall have an F rating of not less than the required fire-resistance rating of the wall penetrated.

302.2.2.2 Ducts and air-transfer openings. Penetrations of fire-resistance-rated walls by ducts and air-transfer openings that are not protected with fire dampers shall comply with this section.

302.2.2.3 Dissimilar materials. Noncombustible penetrating items shall not connect to combustible items beyond the point of firestopping unless it can be demonstrated that the fire-resistance integrity of the wall is maintained.

302.2.3 Horizontal assemblies. Penetrations of a floor, floor/ceiling assembly or the ceiling membrane of a roof/ceiling assembly shall be protected in accordance with Section 707 of the *Building Code of New York State*. Penetrations permitted by Exceptions 3 and 4 of Section 707.2 of the *Building Code of New York State* shall comply with Sections 302.2.3.1 through 302.2.3.3.

302.2.3.1 Through penetrations. Through penetrations of fire-resistance-rated horizontal assemblies shall comply with Section 302.2.3.1.1 or Section 302.2.3.1.2.

Exceptions:
1. Penetrations by steel, ferrous or copper conduits,

pipes, tubes, vents, concrete, or masonry through a single fire-resistance-rated floor assembly where the annular space is protected with materials that prevent the passage of flame and hot gases sufficient to ignite cotton waste where subjected to ASTM E 119 time temperature fire conditions under a minimum positive pressure differential of 0.01 inch (2.49 Pa) of water at the location of the penetration for the time period equivalent to the fire-resistance rating of the construction penetrated. Penetrating items with a maximum 6-inch (152 mm) nominal diameter shall not be limited to the penetration of a single fire-resistance-rated floor assembly, provided that the area of the penetration does not exceed 144 square inches (92 900 mm^2) in any 100 square feet (9.3 m^2) of floor area.

2. Penetrations in a single concrete floor by steel, ferrous or copper conduits, pipes, tubes and vents with a maximum 6-inch (152 mm) nominal diameter, provided that concrete, grout or mortar is installed the full thickness of the floor or the thickness required to maintain the fire-resistance rating. The penetrating items with a maximum 6-inch (152 mm) nominal diameter shall not be limited to the penetration of a single concrete floor, provided that the area of the penetration does not exceed 144 square inches (92 900 mm^2).

3. Electrical outlet boxes of any material are permitted provided that such boxes are tested for use in fire-resistance-rated assemblies and are installed in accordance with the tested assembly.

302.2.3.1.1 Fire-resistance-rated assemblies. Penetrations shall be installed as tested in the approved fire-resistance-rated assembly.

302.2.3.1.2 Through-penetration firestop system. Through penetrations shall be protected by an approved through-penetration firestop system installed and tested in accordance with ASTM E 814, with a minimum positive pressure differential of 0.01 inch (2.49 Pa) of water. The system shall have an F rating and a T rating of not less than 1 hour but not less than the required rating of the floor penetrated.

Exception: Floor penetrations contained and located within the cavity of a wall do not require a T rating.

302.2.3.2 Nonfire-resistance-rated assemblies. Penetrations of horizontal assemblies without a required fire-resistance rating shall meet the requirements of Section 707 of the *Building Code of New York State* or shall comply with Sections 302.2.3.2.1 through 302.2.3.2.2.

302.2.3.2.1 Noncombustible penetrating items. Noncombustible penetrating items that connect not more than three stories are permitted, provided that the annular space is filled with an approved noncombustible material to resist the free passage of flame and the products of combustion.

302.2.3.2.2 Penetrating items. Penetrating items that connect not more than two stories are permitted, provided that the annular space is filled with an approved material to resist the free passage of flame and the products of combustion.

302.2.3.3 Ducts and air-transfer openings. Penetrations of horizontal assemblies by ducts and air-transfer openings that are not required to have dampers shall comply with this section. Ducts and air-transfer openings that are protected with dampers shall comply with Section 715 of the *Building Code of New York State*.

302.2.3.4 Dissimilar materials. Noncombustible penetrating items shall not connect to combustible materials beyond the point of firestopping unless it can be demonstrated that the fire-resistance integrity of the horizontal assembly is maintained.

302.3 Cutting, notching and boring in wood members. The cutting, notching and boring of wood members shall comply with Sections 302.3.1 through 302.3.3 and Appendix E.

302.3.1 Joist notching. Notching at the ends of joists shall not exceed one-fourth the joist depth. Holes bored in joists shall not be within 2 inches (51 mm) of the top and bottom of the joist and their diameter shall not exceed one-third the depth of the member. Notches in the top or bottom of the joist shall not exceed one-sixth the depth and shall not be located in the middle one-third of the span.

302.3.2 Stud cutting and notching. In exterior walls and bearing partitions, any wood stud is permitted to be cut or notched to a depth not exceeding 25 percent of its width. Cutting or notching of studs to a depth not greater than 40 percent of the width of the stud is permitted in nonbearing partitions supporting no loads other than the weight of the partition.

302.3.3 Bored holes. A hole not greater in diameter than 40 percent of the stud depth is permitted to be bored in any wood stud. Bored holes not greater than 60 percent of the depth of the stud are permitted in nonbearing partitions or in any wall where each bored stud is doubled, provided not more than two such successive doubled studs are so bored. In no case shall the edge of the bored hole be nearer than 5/8 inch (15.9 mm) to the edge of the stud. Bored holes shall not be located at the same section of stud as a cut or notch.

302.4 Cutting, notching and boring holes in structural steel framing. The cutting, notching and boring of holes in structural steel framing members shall be as prescribed by the registered design professional.

302.5 Cutting, notching and boring holes in cold-formed steel framing. Flanges and lips of load-bearing, cold-formed steel framing members shall not be cut or notched. Holes in webs of load-bearing, cold-formed steel framing members shall be permitted along the centerline of the web of the framing member and shall not exceed the dimensional limitations, penetration spacing or minimum hole edge distance as prescribed by the registered design professional. Cutting, notching and boring holes of steel floor/roof decking shall be as prescribed by the registered design professional.

302.6 Cutting, notching and boring holes in nonstructural cold-formed steel wall framing. Flanges and lips of nonstructural cold-formed steel wall studs shall be permitted along the centerline of the web of the framing member, shall not exceed 1 1/2 inches (38 mm) in width or 4 inches (102 mm) in length, and the holes shall not be spaced less than 24 inches (610 mm) center to center from another hole or less than 10 inches (254 mm) from the bearing end.

302.7 Alterations to trusses. Truss members and components shall not be cut, drilled, notched, spliced or otherwise altered in any way without the written concurrence and approval of a registered design professional. Alterations resulting in the addition of loads to any member (e.g., HVAC equipment, water heater), shall not be permitted without verification that the truss is capable of supporting such additional loading.

SECTION 303
APPLIANCE LOCATION

303.1 General. Appliances shall be located as required by this section, specific requirements elsewhere in this code and the conditions of the equipment and appliance listing.

303.2 Hazardous locations. Appliances shall not be located in a hazardous location unless listed and approved for the specific installation.

303.3 Prohibited locations. Appliances shall not be located in, or obtain combustion air from, any of the following rooms or spaces:

1. Sleeping rooms.
2. Bathrooms.
3. Toilet rooms.
4. Storage closets.
5. Surgical rooms.

Exceptions:

1. Direct-vent appliances that obtain all combustion air directly from the outdoors.
2. Vented room heaters, wall furnaces, vented decorative appliances and decorative appliances for installation in vented solid fuel-burning fireplaces, provided that the room is not a confined space and the building is not of unusually tight construction.
3. A single wall-mounted unvented room heater equipped with an oxygen depletion safety shutoff system and installed in a bathroom provided that the input rating does not exceed 6000 Btu per hour (1.76 kW) and the bathroom is not a confined space.
4. A single wall-mounted unvented room heater equipped with an oxygen depletion safety shutoff system and installed in a bedroom provided that the input rating does not exceed 10,000 Btu per hour (2.93 kW) and the bedroom is not a confined space.
5. Appliances installed in an enclosure in which all combustion air is taken from the outdoors, in accordance with Section 304.11. Access to such enclosure shall be through a solid weather-stripped door, equipped with an approved self-closing device.

303.4 Protection from physical damage. Appliances shall not be installed in a location where subject to physical damage unless protected by approved barriers meeting the requirements of the *Fire Code of New York State*.

303.5 Indoor locations. Furnaces and boilers installed in closets and alcoves shall be listed for such installation.

303.6 Outdoor locations. Equipment installed in outdoor locations shall be either listed for outdoor installation or provided with protection from outdoor environmental factors that influence the operability, durability, and safety of the equipment.

303.7 Pit locations. Appliances installed in pits or excavations shall not come in direct contact with the surrounding soil. The sides of the pit or excavation shall be held back a minimum of 12 inches (305 mm) from the appliance. Where the depth exceeds 12 inches (305 mm) below adjoining grade, the walls of the pit or excavation shall be lined with concrete or masonry, such concrete or masonry shall extend a minimum of 4 inches (102 mm) above adjoining grade and shall have sufficient lateral load-bearing capacity to resist collapse. The appliance shall be protected from flooding in an approved manner.

SECTION 304
COMBUSTION, VENTILATION, AND DILUTION AIR

304.1 General. The provisions of Section 304 shall apply to gas utilization equipment installed in buildings and requires air for combustion, ventilation, and dilution of flue gases.

Exceptions:
1. Direct vent equipment that is constructed and installed so that all air for combustion is obtained directly from the outdoors and all flue gases are discharged to the outdoors.

2. Enclosed furnaces that incorporate an integral total enclosure and use only outdoor air for combustion and dilution of flue gases.

304.2 Appliance/equipment location. Equipment shall be located so as not to interfere with proper circulation of combustion, ventilation, and dilution air.

304.3 Outdoor air required. Where normal infiltration does not provide the necessary air, outdoor air shall be introduced in accordance with Section 304.11 or 304.13.

304.4 Process air. In addition to air needed for combustion, process air shall be provided as required for cooling of equipment or material, controlling dew point, heating, drying, oxidation, dilution, safety exhaust, odor control, and air for compressors.

304.5 Ventilation air. In addition to air needed for combustion, air shall be supplied for ventilation, including all air required for comfort and proper working conditions for personnel.

304.6 Draft hood/regulator location. A draft hood or a barometric draft regulator shall be installed in the same room or enclosure as the equipment served so as to prevent any difference in pressure between the hood or regulator and the combustion air supply.

304.7 Makeup air provisions. Air requirements for the operation of exhaust fans, kitchen ventilation systems, clothes dryers, and fireplaces shall be considered in determining the adequacy of a space to provide combustion air requirements.

304.8 Combustion air methods. Air for combustion, ventilation, and dilution of flue gases for gas utilization equipment vented by natural draft shall be obtained by application of one of the methods covered in Sections 304.10 through 304.13.

304.9 Unusually tight construction. Equipment located in buildings of unusually tight construction (see definitions in Section 202) shall be provided with air for combustion, ventilation, and dilution of flue gases using one of the methods described in Section 304.11 or 304.13.

304.10 All air from inside the building. A confined space shall be provided with two permanent openings communicating directly with other spaces of sufficient volume so that the combined volume of all spaces meets the criteria for an unconfined space. The total input of all equipment installed in the combined spaces shall be used to determine the required minimum volume. Each opening shall have a minimum free area of not less than 1 square inch per 1,000 Btu per hour (22 cm^2 per kw) of the total input rating of all gas utilization equipment in the confined space, but not less than 100 square inches (64 415 mm^2). One opening shall commence within 12 inches (305 mm) of the top, and one opening shall commence within 12 inches (305 mm) of the bottom, of the enclosure (see Figure 304.10). The minimum dimension of air openings shall be not less than 3 inches (76 mm).

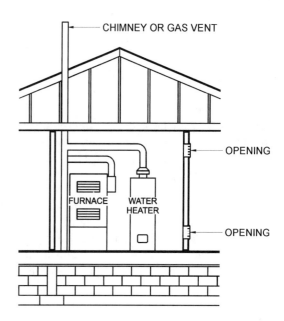

CHIMNEY OR GAS VENT

OPENING

OPENING

FURNACE   WATER HEATER

FIGURE 304.10
APPLIANCES LOCATED IN CONFINED SPACES; ALL AIR FROM INSIDE THE BUILDING (See Section 304.10)

304.11 All air from outdoors. The confined space shall communicate with the outdoors in accordance with Section 304.11.1 or 304.11.2. The minimum dimension of air openings shall not be less than 3 inches (76 mm). Where ducts are used, they shall be of the same cross-sectional area as the free area of the openings to which they connect.

304.11.1 Two opening method. Two permanent openings, one commencing within 12 inches (305 mm) of the top, and one commencing within 12 inches (305 mm) of the bottom, of the enclosure shall be provided. The openings shall communicate directly, or by ducts, with the outdoors or spaces that freely communicate with the outdoors.

Where directly communicating with the outdoors, or where communicating with the outdoors through vertical ducts, each opening shall have a minimum free area of 1 square inch per 4,000 Btu per hour (5.5 cm^2 per kw) of total input rating of all equipment in the enclosure [see Figures 304.11(1) and 304.11(2)].

Where communicating with the outdoors through horizontal ducts, each opening shall have a minimum free area of not less than 1 square inch per 2,000 Btu per hour (11 cm^2 per kw) of total input rating of all equipment in the enclosure [see Figure 304.11(3)].

304.11.2 One opening method. One permanent opening, commencing within 12 inches (305 mm) of the top of the enclosure, shall be provided. The equipment shall have clearances of at least 1 inch (25.4 mm) from the sides and back and 6 inches (152 mm) from the front of the appliance. The opening shall directly communicate with the outdoors or through a vertical or horizontal duct to the outdoors or spaces that freely communicate with the outdoors [see Figure 304.11(4)] and shall have a minimum free area of 1 square inch per 3000 Btu per hr (7 cm^2 per kw) of the total input rating of all equipment located in the enclosure, and not less than the sum of the areas of all vent connectors in the confined space.

304.12 Combination of air from inside and outdoors. Where the building in which the fuel-burning appliances are located is not unusually tight construction and the communicating interior spaces containing the fuel-burning appliances comply with all of the requirements of Section 304.10, except the volumetric requirement of Section 304.10, required combustion and dilution air shall be obtained by opening the room to the outdoors utilizing a combination of inside and outdoor air prorated in accordance with Section 304.12.6. Openings connecting the interior spaces shall comply with Section 304.10. The ratio of interior spaces shall comply with Section 304.12.5. The number, location and ratios of openings connecting the space with the outdoor air shall comply with Sections 304.12.1 through 304.12.4.

304.12.1 Number and location of openings. At least two openings shall be provided, one within 1 foot (305 mm) of the ceiling of the room and one within 1 foot (305 mm) of the floor.

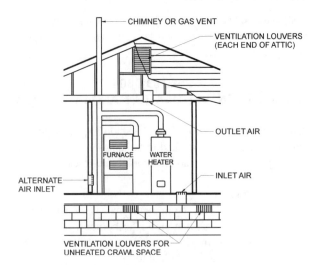

FIGURE 304.11(1)
APPLIANCES LOCATED IN CONFINED SPACES; ALL AIR FROM OUTDOORS—INLET AIR FROM VENTILATED CRAWL SPACE AND OUTLET AIR TO VENTILATED ATTIC (See Section 304.11.1)

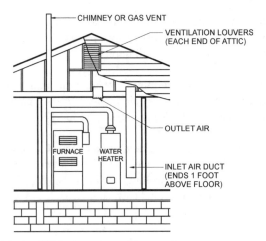

For SI: 1 foot = 304.8 mm.

FIGURE 304.11(2)
APPLIANCES LOCATED IN CONFINED SPACES; ALL AIR FROM OUTDOORS THROUGH VENTILATED ATTIC (See Section 304.11.1)

304.12.2 Ratio of direct openings. Where direct openings to the outdoors are provided in accordance with Section 304.11.1, the ratio of direct openings shall be the sum of the net free areas of both direct openings to the outdoors, divided by the sum of the required areas for both such openings as determined in accordance with Section 304.11.1.

304.12.3 Ratio of horizontal openings. Where openings connected to the outdoors through horizontal ducts are provided in accordance with Section 304.11.1, the ratio of horizontal openings shall be the sum of the net free areas of both such openings, divided by the sum of the required areas for both such openings as determined in accordance with Section 304.11.1.

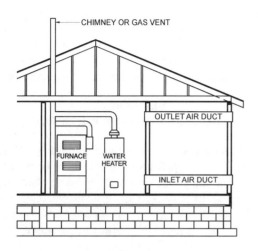

FIGURE 304.11(3)
APPLIANCES LOCATED IN CONFINED SPACES; ALL AIR FROM OUTDOORS (See Section 304.11.1)

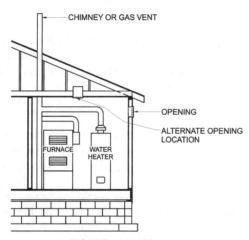

FIGURE 304.11(4)
APPLIANCES LOCATED IN CONFINED SPACES; SINGLE COMBUSTION AIR OPENING, ALL AIR FROM THE OUTDOORS (See Section 304.11.2)

304.12.4 Ratio of vertical openings. Where openings connected to the outdoors through vertical ducts are provided in accordance with Section 304.11.1, the ratio of vertical openings shall be the sum of the net free areas of both such openings, divided by the sum of the required areas for both such openings as determined in accordance with Section 304.11.1.

304.12.5 Ratio of interior spaces. The ratio of interior spaces shall be the available volume of all communicating spaces, divided by the required volume as determined in accordance with Section 304.10.

304.12.6 Prorating of inside and outdoor air. In spaces that utilize a combination of inside and outdoor air, the sum of the ratios of all direct openings, horizontal openings, vertical openings and interior spaces shall equal or exceed 1.

304.13 Specially engineered installations. As an alternative to the provisions of Sections 304.10, 304.11 and 304.12, the necessary supply of air for combustion, ventilation and dilution of flue gases shall be provided by an approved engineered system.

304.14 Louvers and grilles. In calculating free area in Sections 304.10, 304.11 and 304.12, the required size of openings for combustion, ventilation and dilution air shall be based on the net free area of each opening. If the free area through a design of louver or grille is known, it shall be used in calculating the size opening required to provide the free area specified. If the design and free area are not known, it shall be assumed that wood louvers will have 20-25 percent free area and metal louvers and grilles will have 60-75 percent free area. Louvers and grilles shall be fixed in the open position.

Exception: Louvers interlocked with the equipment so that they are proven to be in the full open position prior to main burner ignition and during main burner operation. Means shall be provided to prevent the main burner from igniting if the louvers fail to open during burner start-up and to shut down the main burner if the louvers close during operation.

304.15 Combustion air ducts. Combustion air ducts shall comply with all of the following:

1. Ducts shall be of galvanized steel complying with Chapter 6 of the *Mechanical Code of New York State* or of equivalent corrosion-resistant material approved for this application.

   Exception: Within dwellings units, unobstructed stud and joist spaces shall not be prohibited from conveying combustion air, provided that not more than one required fireblock is removed.

2. Ducts shall terminate in an unobstructed space allowing free movement of combustion air to the appliances.

3. Ducts shall serve a single enclosure.

4. Ducts shall not serve both upper and lower combustion air openings where both such openings are used. The separation between ducts serving upper and lower combustion air openings shall be maintained to the source of combustion air.

5. Ducts shall not be screened where terminating in an attic space.

6. Horizontal upper combustion air ducts shall not slope downward toward the source of combustion air.

SECTION 305
INSTALLATION

305.1 General. Equipment and appliances shall be installed as required by the terms of their approval. Equipment and appliances shall be installed in accordance with the conditions of listing, the manufacturer's installation instructions, and this code. Manufacturers' installation instructions shall be available on the job site at the time of inspection.

Unlisted appliances approved in accordance with Section 301.3 shall be limited to uses recommended by the manufacturer and shall be installed in accordance with the manufacturer's installation instructions and the provisions of this code.

305.2 Elevation of ignition source. Equipment and appliances having an ignition source shall be elevated such that the source of ignition is not less than 18 inches (457 mm) above the floor in hazardous locations and public garages, private garages, repair garages, automotive service stations and parking garages. Such equipment and appliances shall not be installed in Group H occupancies or control areas where open use, handling or dispensing of combustible, flammable or explosive materials occurs. For the purpose of this section, rooms or spaces that are not part of the living space of a dwelling unit and that communicate directly with a private garage through openings shall be considered to be part of the private garage.

305.3 Public garages. Appliances located in public garages, service stations, repair garages or other areas frequented by motor vehicles shall be installed a minimum of 8 feet (2438 mm) above the floor. Where motor vehicles exceed 6 feet (1829 mm) in height and are capable of passing under an appliance, appliances shall be installed a minimum of 2 feet (610 mm) higher above the floor than the height of the tallest vehicle.

> **Exception:** The requirements of this section shall not apply where the appliances are protected from motor vehicle impact and installed in accordance with Section 305.2 and NFPA 88B.

305.4 Private garages. Appliances located in private garages shall be installed with a minimum clearance of 6 feet (1829 mm) above the floor.

> **Exception:** The requirements of this section shall not apply where the appliances are protected from motor vehicle impact and installed in accordance with Section 305.2.

305.5 Construction and protection. Boiler rooms and furnace rooms shall be protected as required by the *Building Code of New York State*.

305.6 Clearances from grade. Equipment and appliances installed at grade level shall be supported on a level concrete slab or other approved material extending above adjoining grade or shall be suspended a minimum of 6 inches (152 mm) above adjoining grade.

305.7 Clearances to combustible construction. Heat-producing equipment and appliances shall be installed to maintain the required clearances to combustible construction as specified in the listing and manufacturer's instructions. Such

clearances shall be reduced only in accordance with Section 308. Clearances to combustibles shall include such considerations as door swing, drawer pull, overhead projections or shelving and window swing. Devices such as door stops or limits and closers shall not be used to provide the required clearances.

SECTION 306
ACCESS AND SERVICE SPACE

306.1 Clearances for maintenance and replacement. Clearances around appliances to elements of permanent construction, including other installed appliances, shall be sufficient to allow inspection, service, repair or replacement without removing such elements of permanent construction or disabling the function of a required fire-resistance-rated assembly.

306.2 Appliances in rooms. Rooms containing appliances requiring access shall be provided with a door and an unobstructed passageway measuring not less than 35 inches (889 mm) wide and 80 inches (2032 mm) high.

> **Exception:** Within a dwelling unit, appliances installed in a compartment, alcove, basement or similar space shall be provided with access by an opening or door and an unobstructed passageway measuring not less than 24 inches (610 mm) wide and large enough to allow removal of the largest appliance in the space, provided that a level service space of not less than 30 inches (762 mm) deep and the height of the appliance, but not less than 30 inches (762 mm), is present at the front or service side of the appliance with the door open.

306.3 Appliances in attics. Attics containing appliances requiring access shall be provided with an opening and unobstructed passageway large enough to allow removal of the largest component of the appliance. The passageway shall not be less than 30 inches (762 mm) high and 22 inches (559 mm) wide and not more than 20 feet (6096 mm) in length when measured along the centerline of the passageway from the opening to the equipment. The passageway shall have continuous solid flooring not less than 24 inches (610 mm) wide. A level service space not less than 30 inches (762 mm) deep and 30 inches (762 mm) wide shall be present at the front or service side of the equipment. The clear access opening dimensions shall be a minimum of 20 inches by 30 inches (508 mm by 762 mm), where such dimensions are large enough to allow removal of the largest component of the appliance.

> **Exception:** The passageway and level service space are not required where the appliance is capable of being serviced and removed through the required opening.

306.3.1 Electrical requirements. A lighting fixture controlled by a switch located at the required passageway opening and a receptacle outlet shall be provided at or near the equipment location in accordance with Chapter 27 of the *Building Code of New York State*.

306.4 Appliances under floors. Underfloor spaces containing appliances requiring access shall be provided with an access opening and unobstructed passageway large enough to remove the largest component of the appliance. The passageway shall not be less than 30 inches (762 mm) high and 22 inches (559 mm) wide, nor more than 20 feet (6096 mm) in length when measured along the centerline of the passageway from the opening to the equipment. A level service space not less than 30 inches (762 mm) deep and 30 inches (762 mm) deep and 30 inches (762 mm) wide shall be present at the front or service side of the appliance. If the depth of the passageway or the service space exceeds 12 inches (305 mm) below the adjoining grade, the walls of the passageway shall be lined with concrete or masonry extending 4 inches (102 mm) above the adjoining grade and having sufficient lateral-bearing capacity to resist collapse. The clear access opening dimensions shall be a minimum of 22 inches by 30 inches (559 mm by 762 mm), where such dimensions are large enough to allow removal of the largest component of the appliance.

> **Exception:** The passageway is not required where the level service space is present when the access is open and the appliance is capable of being serviced and removed through the required opening.

306.4.1 Electrical requirements. A lighting fixture controlled by a switch located at the required passageway opening and a receptacle outlet shall be provided at or near the equipment location in accordance with Chapter 27 of the *Building Code of New York State*.

306.5 Appliances on roofs or elevated structures. Where appliances requiring access are installed on roofs or elevated structures at a height exceeding 16 feet (4877 mm), such access shall be provided by a permanent approved means of access, the extent of which shall be from grade or floor level to the appliance's level service space. Such access shall not require climbing over obstructions greater than 30 inches (762 mm) high or walking on roofs having a slope greater than 4 units vertical in 12 units horizontal (33-percent slope).

306.5.1 Sloped roofs. Where appliances are installed on a roof having a slope of 3 units vertical in 12 units horizontal (25-percent slope) or greater and having an edge more than 30 inches (762 mm) above grade at such edge, a level platform shall be provided on each side of the appliance to which access is required by the manufacturer's installation instructions for service, repair or maintenance. The platform shall not be less than 30 inches (762 mm) in any dimension and shall be provided with guards in accordance with Section 306.6.

306.5.2 Electrical requirements. A receptacle outlet shall be provided at or near the equipment location in accordance with Chapter 27 of the *Building Code of New York State*.

306.6 Guards. Guards shall be provided where appliances, fans or other components that require service are located within 10 feet (3048 mm) of a roof edge or open side of a walking surface and such edge or open side is located more than 30 inches (762 mm) above the floor, roof or grade below. The top of the guard shall be located not less than 42 inches (1067 mm) above the elevated surface adjacent to the guard. The guard shall be constructed so as to prevent the passage of a 21-inch-diameter (533 mm) sphere and shall comply with the loading requirements for guards specified in the *Building Code of New York State*.

SECTION 307
CONDENSATE DISPOSAL

307.1 Fuel-burning appliances. Liquid combustion by-products of condensing appliances shall be collected and discharged to an approved plumbing fixture or disposal area in accordance with the manufacturer's installation instructions. Condensate piping shall be of approved corrosion-resistant material and shall not be smaller than the drain connection on the appliance. Such piping shall maintain a minimum slope in the direction of discharge of not less than one-eighth unit vertical in 12 units horizontal (1-percent slope).

307.2 Drain pipe materials and sizes. Components of the condensate disposal system shall be cast iron, galvanized steel, copper, polyethylene, ABS, CPVC or PVC pipe or tubing. All components shall be selected for the pressure and temperature rating of the installation. Condensate waste and drain line size shall be not less than 3/4-inch internal diameter (19 mm) and shall not decrease in size from the drain connection to the place of condensate disposal. Where the drain pipes from more than one unit are manifolded together for condensate drainage, the pipe or tubing shall be sized in accordance with an approved method. All horizontal sections of drain piping shall be installed in uniform alignment at a uniform slope.

307.3 Traps. Condensate drains shall be trapped as required by the equipment or appliance manufacturer.

SECTION 308
CLEARANCE REDUCTION

308.1 Scope. This section shall govern the reduction in required clearances to combustible materials and combustible assemblies for chimneys, vents, kitchen exhaust equipment, fuel gas appliances, and fuel gas devices and equipment. Clearance requirements for air-conditioning equipment and central heating boilers and furnaces shall comply with Sections 308.3 and 308.4.

308.2 Reduction table. The allowable clearance reduction shall be based on one of the methods specified in Table 308.2 or shall utilize an assembly listed for such application. Where required clearances are not listed in Table 308.2, the reduced clearances shall be determined by linear interpolation between the distances listed in the table. Reduced clearances shall not be derived by extrapolation below the range of the table. The reduction of the required clearances to combustibles for listed and labeled appliances and equipment shall be in accordance with the requirements of this section except that such clearances shall not be reduced where reduction is specifically prohibited by the terms of the appliance or equipment listing [see Figures 308.2(1) through 308.2(3)].

308.3 Clearances for indoor air-conditioning equipment. Clearance requirements for indoor air-conditioning equipment shall comply with Sections 308.3.1 through 308.3.6.

308.3.1 Equipment installed in rooms that are large in comparison with the size of the equipment. Air-conditioning equipment installed in rooms that are large in comparison with the size of the equipment shall be installed with clearances per the terms of their listing and the manufacturer's instructions.

308.3.2 Equipment installed in rooms that are not large in comparison with the size of the equipment. Air-conditioning equipment installed in rooms that are not large in comparison with the size of the equipment, such as alcoves and closets, shall be listed for such installations and installed in accordance with the manufacturer's instructions. Listed clearances shall not be reduced by the protection methods described in Table 308.2, regardless of whether the enclosure is of combustible or noncombustible material.

308.3.3 Clearance reduction. Air-conditioning equipment installed in rooms that are large in comparison with the size of the equipment shall be permitted to be installed with reduced clearances to combustible material provided the combustible material or equipment is protected as described in Table 308.2.

308.3.4 Plenum clearances. Where the plenum is adjacent to plaster on metal lath or noncombustible material attached to combustible material, the clearance shall be measured to the surface of the plaster or other noncombustible finish where the clearance specified is 2 inches (51 mm) or less.

308.3.5 Clearance from supply ducts. Air-conditioning equipment shall have the clearance from supply ducts within 3 feet (914 mm) of the plenum be not less than that specified from the plenum. No clearance is necessary beyond this distance.

308.4 Central-heating boilers and furnaces. Clearance requirements for central-heating boilers and furnaces shall comply with Sections 308.4.1 through 308.4.7. The clearance to this equipment shall not interfere with combustion air, draft hood clearance and relief, and accessibility for servicing.

308.4.1 Equipment installed in rooms that are large in comparison with the size of the equipment. Central-heating furnaces and low-pressure boilers installed in rooms large in comparison with the size of the equipment shall be installed with clearances per terms of their listing and the manufacturer's instructions.

308.4.2 Equipment installed in rooms that are not large in comparison with the size of the equipment. Central-heating furnaces and low-pressure boilers installed in rooms that are not large in comparison with the size of the equipment, such as alcoves and closets, shall be listed for such installations. Listed clearances shall not be reduced by the protection methods described in Table 308.2 and illustrated in Figures 308.2(1) through 308.2(3), regardless of whether the enclosure is of combustible or noncombustible material.

308.4.3 Clearance reduction. Central-heating furnaces and low-pressure boilers installed in rooms that are large in comparison with the size of the equipment shall be permitted to be installed with reduced clearances to combustible material provided the combustible material or equipment is protected as described in Table 308.2.

308.4.4 Clearance for servicing equipment. Front clearance shall be sufficient for servicing the burner and the furnace or boiler.

308.4.5 Plenum clearances. Where the plenum is adjacent to plaster on metal lath or noncombustible material attached to combustible material, the clearance shall be measured to the surface of the plaster or other noncombustible finish where the clearance specified is 2 inches (51 mm) or less.

308.4.6 Clearance from supply ducts. Central-heating furnaces shall have the clearance from supply ducts within 3 feet (914 mm) of the plenum be not less than that specified from the plenum. No clearance is necessary beyond this distance.

308.4.7 Other central heating furnaces. Central heating furnaces other than those listed in Section 308.4.6 or 308.4.7 shall have clearances from the supply ducts of not less than 18 inches (457 mm) from the plenum for the first 3 feet (914 mm), then 6 inches (152 mm) for the next 3 feet (914 mm) and 1 inch (25 mm) beyond 6 feet (1829 mm).

SECTION 309
ELECTRICAL

309.1 Grounding. Gas piping shall not be used as a grounding electrode.

309.2 Connections. Electrical connections between equipment and the building wiring, including the grounding of the equipment, shall conform to Chapter 27 of the *Building Code of New York State*.

TABLE 308.2[a-k]
REDUCTION OF CLEARANCES WITH SPECIFIED FORMS OF PROTECTION

TYPE OF PROTECTION APPLIED TO AND COVERING ALL SURFACES OF COMBUSTIBLE MATERIAL WITHIN THE DISTANCE SPECIFIED AS THE REQUIRED CLEARANCE WITH NO PROTECTION [see Figures 308.2(1), 308.2(2), and 308.2(3)]	WHERE THE REQUIRED CLEARANCE WITH NO PROTECTION FROM APPLIANCE, VENT CONNECTOR, OR SINGLE WALL METAL PIPE IS (INCHES):									
	36		18		12		9		6	
	Allowable clearances with specified protection (inches) Use Column 1 for clearances above appliance or horizontal connector. Use Column 2 for clearances from appliance, vertical connector, and single-wall metal pipe.									
	Above Col. 1	Sides and rear Col. 2	Above Col. 1	Sides and rear Col. 2	Above Col. 1	Sides and rear Col. 2	Above Col. 1	Sides and rear Col. 2	Above Col. 1	Sides and rear Col. 2
1. 3 1/2-inch thick masonry wall without ventilated airspace	—	24	—	12	—	9	—	6	—	5
2. 1/2-inch insulation board over 1-inch glass fiber or mineral wool batts	24	18	12	9	9	6	6	5	4	3
3. 0.024 sheet metal over 1-inch glass fiber or mineral wool batts reinforced with wire on rear face with ventilated airspace	18	12	9	6	6	4	5	3	3	3
4. 3 1/2-inch thick masonry wall with ventilated airspace	—	12	—	6	—	6	—	6	—	6
5. 0.024 sheet metal with ventilated airspace	18	12	9	6	6	4	5	3	3	2
6. 1/2-inch thick insulation board with ventilated airspace	18	12	9	6	6	4	5	3	3	3
7. 0.024 sheet metal with ventilated airspace over 0.024 sheet metal with ventilated airspace	18	12	9	6	6	4	5	3	3	3
8. 1-inch glass fiber or mineral wool batts sandwiched between two sheets 0.024 sheet metal with ventilated airspace	18	12	9	6	6	4	5	3	3	3

For SI: 1 inch = 25.4 mm, °C = [(°F-32)/1.8], 1 pound per cubic foot = 16.02 kg/m^3, 1 Btu per inch per square foot per hour per °F = 0.144 w/m^2.

a. Reduction of clearances from combustible materials shall not interfere with combustion air, draft hood clearance and relief, and accessibility of servicing.
b. All clearances shall be measured from the outer surface of the combustible material to the nearest point on the surface of the appliance, disregarding any intervening protection applied to the combustible material.
c. Spacers and ties shall be of noncombustible material. No spacer or tie shall be used directly opposite appliance or connector.
d. For all clearance reduction systems using a ventilated airspace, adequate provision for air circulation shall be provided as described [see Figures 308.2(2) and 308.2(3)].
e. There shall be at least 1 inch between clearance reduction systems and combustible walls and ceilings for reduction systems using ventilated airspace.
f. If a wall protector is mounted on a single flat wall away from corners, adequate air circulation shall be permitted to be provided by leaving only the bottom and top edges or only the side and top edges open with at least 1 inch air gap.
g. Mineral wool batts (blanket or board) shall have a minimum density of 8 pounds per cubic foot and a minimum melting point of 1500°F.
h. Insulation material used as part of clearance reduction system shall have a thermal conductivity of 1.0 Btu per inch per square foot per hour per °F or less.
i. There shall be at least 1 inch between the appliance and the protector. In no case shall the clearance between the appliance and the combustible surface be reduced below that allowed in Table 308.2.
j. All clearances and thicknesses are minimum; larger clearances and thicknesses are acceptable.
k. Listed single-wall connectors shall be permitted to be installed in accordance with the terms of their listing and the manufacturer's instructions.

FIGURE 308.2(1) – FIGURE 308.2(3)

GENERAL REGULATIONS

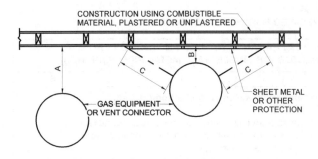

"A" equals the reduced clearance with no protection.

"B" equals the reduced clearance permitted in accordance with Table 308.2. The protection applied to the construction using combustible material shall extend far enough in each direction to make "C" equal to "A."

FIGURE 308.2(1)
EXTENT OF PROTECTION NECESSARY
TO REDUCE CLEARANCES FROM GAS EQUIPMENT
OR VENT CONNECTIONS

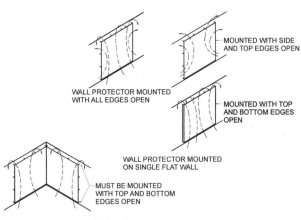

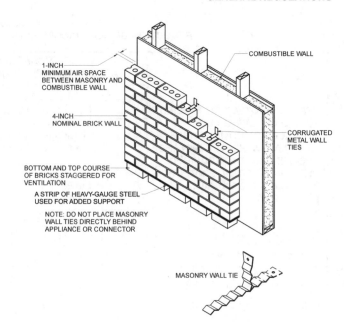

For SI: 1 inch = 25.4 mm.

FIGURE 308.2(3)
EXTENT OF PROTECTION NECESSARY
TO REDUCE CLEARANCES FROM GAS EQUIPMENT
OR VENT CONNECTIONS

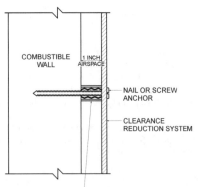

1-INCH NONCOMBUSTIBLE SPACER SUCH AS STACKED WASHERS, SMALL-DIAMETER PIPE, TUBING OR ELECTRICAL CONDUIT.

MASONRY WALLS CAN BE ATTACHED TO COMBUSTIBLE WALLS USING WALL TIES.

DO NOT USE SPACERS DIRECTLY BEHIND APPLIANCE OR CONNECTOR.

For SI: 1 inch = 25.4 mm.

FIGURE 308.2(2)
WALL PROTECTOR CLEARANCE
REDUCTION SYSTEM

CHAPTER 4
GAS PIPING INSTALLATIONS

SECTION 401
GENERAL

401.1 Scope. This chapter shall govern the design, installation, modification and maintenance of piping systems. The applicability of this code to piping systems extends from the point of delivery to the connections with the equipment and includes the design, materials, components, fabrication, assembly, installation, testing, inspection, operation and maintenance of such piping systems.

401.1.1 Utility piping systems located within buildings. Utility service piping located within buildings shall be installed in accordance with the structural safety and fire protection provisions of the *Building Code of New York State*.

401.2 Liquefied petroleum gas storage. The storage system for liquefied petroleum gas shall be designed and installed in accordance with the *Fire Code of New York State* and NFPA 58.

401.3 Modifications to existing systems. In modifying or adding to existing piping systems, sizes shall be maintained in accordance with this chapter.

401.4 Additional appliances. Where an additional appliance is to be served, the existing piping shall be checked to determine if it has adequate capacity for all appliances served. If inadequate, the existing system shall be enlarged as required or separate piping of adequate capacity shall be provided.

401.5 Identification. For other than black steel pipe, exposed piping shall be identified by a yellow label marked "Gas" in black letters. The marking shall be spaced at intervals not exceeding 5 feet (1524 mm). The marking shall not be required on pipe located in the same room as the equipment served.

401.6 Interconnections. Where two or more meters are installed on the same premises but supply separate consumers, the piping systems shall not be interconnected on the outlet side of the meters.

401.7 Piping meter identification. Piping from multiple meter installations shall be marked with an approved permanent identification by the installer so that the piping system supplied by each meter is readily identifiable.

401.8 Minimum sizes. All pipe utilized for the installation, extension and alteration of any piping system shall be sized to supply the full number of outlets for the intended purpose and shall be sized in accordance with Section 402.

SECTION 402
PIPE SIZING

402.1 General considerations. Piping systems shall be of such size and so installed as to provide a supply of gas sufficient to meet the maximum demand without undue loss of pressure between the point of delivery and the gas utilization equipment.

402.2 Maximum gas demand. The volume of gas to be provided, in cubic feet per hour, shall be determined directly from the manufacturers' input ratings of the gas utilization equipment served. Where input rating is not indicated, the gas supplier, equipment manufacturer, or a qualified agency shall be contacted for estimating the volume of gas to be supplied.

The total connected hourly load shall be used as the basis for piping sizing assuming that all equipment could be operating at full capacity simultaneously. Where a diversity of load can be established, pipe sizing shall be permitted to be based on such loads.

402.3 Sizing. Gas piping shall be sized in accordance with Tables 402.3(1) through 402.3(34).

402.4 Allowable pressure drop. The design pressure loss in any piping system under maximum probable flow conditions, from the point of delivery to the inlet connection of the equipment, shall be such that the supply pressure at the equipment is greater than the minimum pressure required for proper equipment operation.

402.5 Maximum design operating pressure. The maximum design operating pressure for piping systems located inside buildings shall not exceed 5 psig (34 kPa gauge) except where one or more of the following conditions are met:

1. The piping system is welded.
2. The piping is located in a ventilated chase or otherwise enclosed for protection against accidental gas accumulation.
3. The piping is located inside buildings or separate areas of buildings used exclusively for:
    3.1. Industrial processing or heating,
    3.2. Research,
    3.3. Warehousing, or
    3.4. Boiler or mechanical equipment rooms.
4. The piping is a temporary installation for buildings under construction.

402.5.1 Liquefied petroleum gas systems. The operating pressure for undiluted LP-gas systems shall not exceed 20 psig (140 kPa gauge). Buildings having systems designed to operate below -5°F (-21°C) or with butane or a propane-butane mix shall be designed to either accommodate liquid LP-gas or prevent LP-gas vapor from condensing into a liquid.

Exception: Buildings, or separate areas of buildings, constructed in accordance with Chapter 7 of NFPA 58, and used exclusively to house industrial processes, research and experimental laboratories, or equipment or processing having similar hazards.

TABLE 402.3(1)
MAXIMUM CAPACITY OF PIPE IN CUBIC FEET OF GAS PER HOUR FOR GAS PRESSURES
OF 0.5 PSI OR LESS AND A PRESSURE DROP OF 0.3-INCH WATER COLUMN
(Based on a 0.60 Specific Gravity Gas)

NOMINAL IRON PIPE SIZE (inches)	INTERNAL DIAMETER (inches)	LENGTH OF PIPE (feet)													
		10	20	30	40	50	60	70	80	90	100	125	150	175	200
1/4	.364	32	22	18	15	14	12	11	11	10	9	8	8	7	6
3/8	.493	72	49	40	34	30	27	25	23	22	21	18	17	15	14
1/2	.622	132	92	73	63	56	50	46	43	40	38	34	31	28	26
3/4	.824	278	190	152	130	115	105	96	90	84	79	72	64	59	55
1	1.049	520	350	285	245	215	195	180	170	160	150	130	120	110	100
1 1/4	1.380	1,050	730	590	500	440	400	370	350	320	305	275	250	225	210
1 1/2	1.610	1,600	1,100	890	760	670	610	560	530	490	460	410	380	350	320
2	2.067	3,050	2,100	1,650	1,450	1,270	1,150	1,050	990	930	870	780	710	650	610
2 1/2	2.469	4,800	3,300	2,700	2,300	2,000	1,850	1,700	1,600	1,500	1,400	1,250	1,130	1,050	980
3	3.068	8,500	5,900	4,700	4,100	3,600	3,250	3,000	2,800	2,600	2,500	2,200	2,000	1,850	1,700
4	4.026	17,500	12,000	9,700	8,300	7,400	6,800	6,200	5,800	5,400	5,100	4,500	4,100	3,800	3,500

For SI: 1 inch = 25.4 mm, 1 foot = 304.8 mm, 1 cubic foot per hour = 0.0283 m³/h, 1 pound per square inch = 6.895 kPa, 1-inch water column = 0.2488 kPa.

TABLE 402.3(2)
MAXIMUM CAPACITY OF PIPE IN CUBIC FEET OF GAS PER HOUR FOR GAS PRESSURES
OF 0.5 PSI OR LESS AND A PRESSURE DROP OF 0.5-INCH WATER COLUMN
(Based on a 0.60 Specific Gravity Gas)

NOMINAL IRON PIPE SIZE (inches)	INTERNAL DIAMETER (inches)	LENGTH OF PIPE (feet)													
		10	20	30	40	50	60	70	80	90	100	125	150	175	200
1/4	.364	43	29	24	20	18	16	15	14	13	12	11	10	9	8
3/8	.493	95	65	52	45	40	36	33	31	29	27	24	22	20	19
1/2	.622	175	120	97	82	73	66	61	57	53	50	44	40	37	35
3/4	.824	360	250	200	170	151	138	125	118	110	103	93	84	77	72
1	1.049	680	465	375	320	285	260	240	220	205	195	175	160	145	135
1 1/4	1.380	1,400	950	770	660	580	530	490	460	430	400	360	325	300	280
1 1/2	1.610	2,100	1,460	1,180	990	900	810	750	690	650	620	550	500	460	430
2	2.067	3,950	2,750	2,200	1,900	1,680	1,520	1,400	1,300	1,220	1,150	1,020	950	850	800
2 1/2	2.469	6,300	4,350	3,520	3,000	2,650	2,400	2,250	2,050	1,950	1,850	1,650	1,500	1,370	1,280
3	3.068	11,000	7,700	6,250	5,300	4,750	4,300	3,900	3,700	3,450	3,250	2,950	2,650	2,450	2,280
4	4.026	23,000	15,800	12,800	10,900	9,700	8,800	8,100	7,500	7,200	6,700	6,000	5,500	5,000	4,600

For SI: 1 inch = 25.4 mm, 1 foot = 304.8 mm, 1 cubic foot per hour = 0.0283m³/h, 1 pound per square inch = 6.895 kPa, 1-inch water column = 0.2488 kPa.

TABLE 402.3(3)
PIPE SIZING TABLE FOR 2 PSI PRESSURE CAPACITY OF PIPES OF DIFFERENT DIAMETERS AND LENGTHS IN CUBIC FEET PER HOUR FOR AN INITIAL PRESSURE OF 2.0 PSI WITH A 1.0 PSI PRESSURE DROP AND A GAS OF 0.6 SPECIFIC GRAVITY

PIPE SIZE OF SCHEDULE 40 STANDARD PIPE (inches)	INTERNAL DIAMETER (inches)	TOTAL EQUIVALENT LENGTH OF PIPE (feet)													
		10	20	30	40	50	60	70	80	90	100	125	150	175	200
1/2	.622	1,506	1,065	869	753	673	615	569	532	502	462	414	372	344	318
3/4	.824	3,041	2,150	1,756	1,521	1,360	1,241	1,150	1,075	1,014	934	836	751	695	642
1	1.049	5,561	3,932	3,211	2,781	2,487	2,270	2,102	1,966	1,854	1,708	1,528	1,373	1,271	1,174
1 1/4	1.380	11,415	8,072	6,591	5,708	5,105	4,660	4,315	4,036	3,805	3,508	3,138	2,817	2,608	2,413
1 1/2	1.610	17,106	12,096	9,876	8,553	7,650	6,983	6,465	6,048	5,702	5,257	4,702	4,222	3,909	3,613
2	2.067	32,944	23,295	19,020	16,472	14,733	13,449	12,452	11,647	10,981	10,125	9,056	8,130	7,527	6,959
2 1/2	2.469	52,505	37,127	30,314	26,253	23,481	21,435	19,845	18,563	17,502	16,138	14,434	12,960	11,999	11,093
3	3.068	92,819	65,633	53,589	46,410	41,510	37,893	35,082	32,817	30,940	28,530	25,518	22,911	21,211	19,608
4	4.026	189,326	133,873	109,307	94,663	84,669	77,292	71,558	66,937	63,109	58,194	52,050	46,732	43,265	39,997

For SI: 1 inch = 25.4 mm, 1 foot = 304.8 mm, 1 cubic foot per hour = 0.0283 m³/h, 1 pound per square inch = 6.895 kPa.

TABLE 402.3(4)
PIPE SIZING TABLE FOR 5 PSI PRESSURE CAPACITY OF PIPES OF DIFFERENT DIAMETERS AND LENGTHS IN CUBIC FEET PER HOUR FOR AN INITIAL PRESSURE OF 5.0 PSI WITH A 3.5 PSI PRESSURE DROP AND A GAS OF 0.6 SPECIFIC GRAVITY

PIPE SIZE OF SCHEDULE 40 STANDARD PIPE (inches)	INTERNAL DIAMETER (inches)	TOTAL EQUIVALENT LENGTH OF PIPE (feet)													
		10	20	30	40	50	60	70	80	90	100	125	150	175	200
1/2	.622	3,185	2,252	1,839	1,593	1,425	1,301	1,204	1,153	1,062	979	876	786	728	673
3/4	.824	6,434	4,550	3,715	3,217	2,878	2,627	2,432	2,330	2,145	1,978	1,769	1,589	1,471	1,360
1	1.049	11,766	8,320	6,793	5,883	5,262	4,804	4,447	4,260	3,922	3,617	3,235	2,905	2,690	2,487
1 1/4	1.380	24,161	17,084	13,949	12,080	10,805	9,864	9,132	8,542	8,054	7,427	6,643	5,964	5,522	5,104
1 1/2	1.610	36,206	25,602	20,904	18,103	16,192	14,781	13,685	12,801	12,069	11,128	9,953	8,937	8,274	7,649
2	2.067	69,727	49,305	40,257	34,864	31,183	28,466	26,354	24,652	23,242	21,433	19,170	17,211	15,934	14,729
2 1/2	2.469	111,133	78,583	64,162	55,566	49,700	45,370	42,004	39,291	37,044	34,159	30,553	27,431	25,396	23,478
3	3.068	196,468	138,924	113,431	98,234	87,863	80,208	74,258	69,462	65,489	60,387	54,012	48,494	44,897	41,504
4	4.026	400,732	283,361	231,363	200,366	179,213	163,598	151,463	141,680	133,577	123,173	110,169	98,911	91,574	84,656

For SI: 1 inch = 25.4 mm, 1 foot = 304.8 mm, 1 cubic foot per hour = 0.0283 m³/h, 1 pound per square inch = 6.895 kPa.

TABLE 402.3(5) – TABLE 402.3(6)

GAS PIPING INSTALLATIONS

TABLE 402.3(5)
PIPE SIZING TABLE FOR PRESSURES UNDER 1 POUND APPROXIMATE CAPACITY OF PIPES
OF DIFFERENT DIAMETERS AND LENGTHS IN CUBIC FEET PER HOUR WITH PRESSURE
DROP OF 0.3-INCH WATER COLUMN AND 0.6 SPECIFIC GRAVITY

PIPE SIZE OF SCHEDULE 40 STANDARD PIPE (inches)	INTERNAL DIAMETER (inches)	TOTAL EQUIVALENT LENGTH OF PIPE (feet)										
		50	100	150	200	250	300	400	500	1000	1500	2000
1.00	1.049	215	148	119	102	90	82	70	62	43	34	29
1.25	1.380	442	304	244	209	185	168	143	127	87	70	60
1.50	1.610	662	455	366	313	277	251	215	191	131	105	90
2.00	2.067	1,275	877	704	602	534	484	414	367	252	203	173
2.50	2.469	2,033	1,397	1,122	960	851	771	660	585	402	323	276
3.00	3.068	3,594	2,470	1,983	1,698	1,505	1,363	1,167	1,034	711	571	488
3.50	3.548	5,262	3,616	2,904	2,485	2,203	1,996	1,708	1,514	1,041	836	715
4.00	4.026	7,330	5,038	4,046	3,462	3,069	2,780	2,380	2,109	1,450	1,164	996
5.00	5.047	13,261	9,114	7,319	6,264	5,552	5,030	4,305	3,816	2,623	2,106	1,802
6.00	6.065	21,472	14,758	11,851	10,143	8,990	8,145	6,971	6,178	4,246	3,410	2,919
8.00	7.981	44,118	30,322	24,350	20,840	18,470	16,735	14,323	12,694	8,725	7,006	5,997
10.00	10.020	80,130	55,073	44,225	37,851	33,547	30,396	26,015	23,056	15,847	12,725	10,891
12.00	11.938	126,855	87,187	70,014	59,923	53,109	48,120	41,185	36,501	25,087	20,146	17,242

For SI: 1 inch = 25.4 mm, 1 foot = 304.8 mm, 1 cubic foot per hour = 0.0283 m³/h, 1 pound per square inch = 6.895 kPa, 1-inch water column = 0.2488 kPa.

TABLE 402.3(6)
PIPE SIZING TABLE FOR PRESSURES UNDER 1 POUND APPROXIMATE CAPACITY OF PIPES
OF DIFFERENT DIAMETERS AND LENGTHS IN CUBIC FEET PER HOUR WITH PRESSURE
DROP OF 0.5-INCH WATER COLUMN AND 0.6 SPECIFIC GRAVITY

PIPE SIZE OF SCHEDULE 40 STANDARD PIPE (inches)	INTERNAL DIAMETER (inches)	TOTAL EQUIVALENT LENGTH OF PIPE (feet)										
		50	100	150	200	250	300	400	500	1000	1500	2000
1.00	1.049	284	195	157	134	119	108	92	82	56	45	39
1.25	1.380	583	400	322	275	244	221	189	168	115	93	79
1.50	1.610	873	600	482	412	366	331	283	251	173	139	119
2.00	2.067	1,681	1,156	928	794	704	638	546	484	333	267	229
2.50	2.469	2,680	1,842	1,479	1,266	1,122	1,017	870	771	530	426	364
3.00	3.068	4,738	3,256	2,615	2,238	1,983	1,797	1,538	1,363	937	752	644
3.50	3.548	6,937	4,767	3,828	3,277	2,904	2,631	2,252	1,996	1372	1,102	943
4.00	4.026	9,663	6,641	5,333	4,565	4,046	3,666	3,137	2,780	1911	1,535	1,313
5.00	5.047	17,482	12,015	9,649	8,258	7,319	6,632	5,676	5,030	3457	2,776	2,376
6.00	6.065	28,308	19,456	15,624	13,372	11,851	10,738	9,190	8,145	5598	4,496	3,848
8.00	7.981	58,161	39,974	32,100	27,474	24,350	22,062	18,883	16,735	11502	9,237	7,905
10.00	10.020	105,636	72,603	58,303	49,900	44,225	40,071	34,296	30,396	20891	16,776	14,358
12.00	11.938	167,236	114,940	92,301	78,998	70,014	63,438	54,295	48,120	33073	26,559	22,731

For SI: 1 inch = 25.4 mm, 1 foot = 304.8 mm, 1 cubic foot per hour = 0.0283 m³/h, 1 pound per square inch = 6.895 kPa, 1-inch water column = 0.2488 kPa.

TABLE 402.3(7)
PIPE SIZING TABLE FOR 1 POUND PRESSURE CAPACITY OF PIPES OF DIFFERENT DIAMETERS AND LENGTHS IN CUBIC FEET PER HOUR FOR AN INITIAL PRESSURE OF 1.0 PSI WITH A 10-PERCENT PRESSURE DROP AND A GAS OF 0.6 SPECIFIC GRAVITY

PIPE SIZE OF SCHEDULE 40 STANDARD PIPE (inches)	INTERNAL DIAMETER (inches)	TOTAL EQUIVALENT LENGTH OF PIPE (feet)										
		50	100	150	200	250	300	400	500	1000	1500	2000
1.00	1.049	717	493	396	338	300	272	233	206	142	114	97
1.25	1.380	1,471	1,011	812	695	616	558	478	423	291	234	200
1.50	1.610	2,204	1,515	1,217	1,041	923	836	716	634	436	350	300
2.00	2.067	4,245	2,918	2,343	2,005	1,777	1,610	1,378	1,222	840	674	577
2.50	2.469	6,766	4,651	3,735	3,196	2,833	2,567	2,197	1,947	1,338	1,075	920
3.00	3.068	11,962	8,221	6,602	5,650	5,008	4,538	3,884	3,442	2,366	1,900	1,626
3.50	3.548	17,514	12,037	9,666	8,273	7,332	6,644	5,686	5,039	3,464	2,781	2,381
4.00	4.026	24,398	16,769	13,466	11,525	10,214	9,255	7,921	7,020	4,825	3,875	3,316
5.00	5.047	44,140	30,337	24,362	20,851	18,479	16,744	14,330	12,701	8,729	7,010	6,000
6.00	6.065	71,473	49,123	39,447	33,762	29,923	27,112	23,204	20,566	14,135	11,351	9,715
8.00	7.981	14,6849	100,929	81,049	69,368	61,479	55,705	47,676	42,254	29,041	23,321	19,960
10.00	10.020	26,6718	183,314	147,207	125,990	111,663	101,175	86,592	76,745	52,747	42,357	36,252
12.00	11.938	42,2248	290,209	233,048	199,459	176,777	160,172	137,087	121,498	83,505	67,057	57,392

For SI: 1 inch = 25.4 mm, 1 foot = 304.8 mm, 1 cubic foot per hour = 0.0283 m³/h, 1 pound per square inch = 6.895 kPa.

TABLE 402.3(8)
PIPE SIZING TABLE FOR 2 POUNDS CAPACITY OF PIPES OF DIFFERENT DIAMETERS AND LENGTHS IN CUBIC FEET PER HOUR FOR AN INITIAL PRESSURE OF 2.0 PSI WITH A 10-PERCENT PRESSURE DROP AND A GAS OF 0.6 SPECIFIC GRAVITY

PIPE SIZE OF SCHEDULE 40 STANDARD PIPE (inches)	INTERNAL DIAMETER (inches)	TOTAL EQUIVALENT LENGTH OF PIPE (feet)										
		50	100	150	200	250	300	400	500	1000	1500	2000
1.00	1.049	1,112	764	614	525	466	422	361	320	220	177	151
1.25	1.380	2,283	1,569	1,260	1,079	956	866	741	657	452	363	310
1.50	1.610	3,421	2,351	1,888	1,616	1,432	1,298	1,111	984	677	543	465
2.00	2.067	6,589	4,528	3,636	3,112	2,758	2,499	2,139	1,896	1,303	1,046	896
2.50	2.469	10,501	7,217	5,796	4,961	4,396	3,983	3,409	3,022	2,077	1,668	1,427
3.00	3.068	18,564	12,759	10,246	8,769	7,772	7,042	6,027	5,342	3,671	2,948	2,523
3.50	3.548	27,181	18,681	15,002	12,840	11,379	10,311	8,825	7,821	5,375	4,317	3,694
4.00	4.026	37,865	26,025	20,899	17,887	15,853	14,364	12,293	10,895	7,488	6,013	5,147
5.00	5.047	68,504	47,082	37,809	32,359	28,680	25,986	22,240	19,711	13,547	10,879	9,311
6.00	6.065	110,924	76,237	61,221	52,397	46,439	42,077	36,012	31,917	21,936	17,616	15,077
8.00	7.981	227,906	156,638	125,786	107,657	95,414	86,452	73,992	65,578	45,071	36,194	30,977
10.00	10.020	413,937	284,497	228,461	195,533	173,297	157,020	134,389	119,106	81,861	65,737	56,263
12.00	11.938	655,315	450,394	361,682	309,553	274,351	248,582	212,754	188,560	129,596	104,070	89,071

For SI: 1 inch = 25.4 mm, 1 foot = 304.8 mm, 1 cubic foot per hour = 0.0283 m³/h, 1 pound per square inch = 6.895 kPa.

TABLE 402.3(9) – TABLE 402.3(10)   GAS PIPING INSTALLATIONS

TABLE 402.3(9)
PIPE SIZING TABLE FOR 5 POUNDS PRESSURE CAPACITY OF PIPES OF DIFFERENT
DIAMETERS AND LENGTHS IN CUBIC FEET PER HOUR FOR AN INITIAL PRESSURE OF 5.0 PSI
WITH A 10-PERCENT PRESSURE DROP AND A GAS OF 0.6 SPECIFIC GRAVITY

PIPE SIZE OF SCHEDULE 40 STANDARD PIPE (inches)	INTERNAL DIAMETER (inches)	TOTAL EQUIVALENT LENGTH OF PIPE (feet)										
		50	100	150	200	250	300	400	500	1000	1500	2000
1.00	1.049	1,989	1,367	1,098	940	833	755	646	572	393	316	270
1.25	1.380	4,084	2,807	2,254	1,929	1,710	1,549	1,326	1,175	808	649	555
1.50	1.610	6,120	4,206	3,378	2,891	2,562	2,321	1,987	1,761	1,210	972	832
2.00	2.067	11,786	8,101	6,505	5,567	4,934	4,471	3,827	3,391	2,331	1,872	1,602
2.50	2.469	18,785	12,911	10,368	8,874	7,865	7,126	6,099	5,405	3,715	2,983	2,553
3.00	3.068	33,209	22,824	18,329	15,687	13,903	12,597	10,782	9,556	6,568	5,274	4,514
3.50	3.548	48,623	33,418	26,836	22,968	20,356	18,444	15,786	13,991	9,616	7,722	6,609
4.00	4.026	67,736	46,555	37,385	31,997	28,358	25,694	21,991	19,490	13,396	10,757	9,207
5.00	5.047	122,544	84,224	67,635	57,887	51,304	46,485	39,785	35,261	24,235	19,461	16,656
6.00	6.065	198,427	136,378	109,516	93,732	83,073	75,270	64,421	57,095	39,241	31,512	26,970
8.00	7.981	407,692	280,204	225,014	192,583	170,683	154,651	132,361	117,309	80,626	64,745	55,414
10.00	10.020	740,477	508,926	408,686	349,782	310,005	280,887	240,403	213,065	146,438	117,595	100,646
12.00	11.938	1,172,269	805,694	647,001	553,749	490,777	444,680	380,588	337,309	231,830	186,168	159,336

For SI: 1 inch = 25.4 mm, 1 foot = 304.8 mm, 1 cubic foot per hour = 0.0283 m³/h, 1 pound per square inch = 6.895 kPa.

TABLE 402.3(10)
PIPE SIZING TABLE FOR 10 POUNDS PRESSURE CAPACITY OF PIPES OF DIFFERENT
DIAMETERS AND LENGTHS IN CUBIC FEET PER HOUR FOR AN INITIAL PRESSURE OF 10.0 PSI
WITH A 10-PERCENT PRESSURE DROP AND A GAS OF 0.6 SPECIFIC GRAVITY

PIPE SIZE OF SCHEDULE 40 STANDARD PIPE (inches)	INTERNAL DIAMETER (inches)	TOTAL EQUIVALENT LENGTH OF PIPE (feet)										
		50	100	150	200	250	300	400	500	1000	1500	2000
1.00	1.049	3,259	2,240	1,798	1,539	1,364	1,236	1,058	938	644	517	443
1.25	1.380	6,690	4,598	3,692	3,160	2,801	2,538	2,172	1,925	1,323	1,062	909
1.50	1.610	10,024	6,889	5,532	4,735	4,197	3,802	3,254	2,884	1,982	1,592	1,362
2.00	2.067	19,305	13,268	10,655	9,119	8,082	7,323	6,268	5,555	3,818	3,066	2,624
2.50	2.469	30,769	21,148	16,982	14,535	12,882	11,672	9,990	8,854	6,085	4,886	4,182
3.00	3.068	54,395	37,385	30,022	25,695	22,773	20,634	17,660	15,652	10,757	8,638	7,393
3.50	3.548	79,642	54,737	43,956	37,621	33,343	30,211	25,857	22,916	15,750	12,648	10,825
4.00	4.026	110,948	76,254	61,235	52,409	46,449	42,086	36,020	31,924	21,941	17,620	15,080
5.00	5.047	200,720	137,954	110,782	94,815	84,033	76,140	65,166	57,755	39,695	31,876	27,282
6.00	6.065	325,013	223,379	179,382	153,527	136,068	123,288	105,518	93,519	64,275	51,615	44,176
8.00	7.981	667,777	458,959	368,561	315,440	279,569	253,310	216,800	192,146	132,061	106,050	90,765
10.00	10.020	1,212,861	833,593	669,404	572,924	507,772	460,078	393,767	348,988	239,858	192,614	164,853
12.00	11.938	1,920,112	1,319,682	1,059,751	907,010	803,866	728,361	623,383	552,493	379,725	304,933	260,983

For SI: 1 inch = 25.4 mm, 1 foot = 304.8 mm, 1 cubic foot per hour = 0.0283 m³/h, 1 pound per square inch = 6.895 kPa.

TABLE 402.3(11)
PIPE SIZING TABLE FOR 20 POUNDS PRESSURE CAPACITY OF PIPES OF DIFFERENT DIAMETERS AND LENGTHS IN CUBIC FEET PER HOUR FOR AN INITIAL PRESSURE OF 20.0 PSI WITH A 10-PERCENT PRESSURE DROP AND A GAS OF 0.6 SPECIFIC GRAVITY

PIPE SIZE OF SCHEDULE 40 STANDARD PIPE (inches)	INTERNAL DIAMETER (inches)	TOTAL EQUIVALENT LENGTH OF PIPE (feet)										
		50	100	150	200	250	300	400	500	1000	1500	2000
1.00	1.049	5,674	3,900	3,132	2,680	2,375	2,152	1,842	1,633	1,122	901	771
1.25	1.380	11,649	8,006	6,429	5,503	4,877	4,419	3,782	3,352	2,304	1,850	1,583
1.50	1.610	17,454	11,996	9,633	8,245	7,307	6,621	5,667	5,022	3,452	2,772	2,372
2.00	2.067	33,615	23,103	18,553	15,879	14,073	12,751	10,913	9,672	6,648	5,338	4,569
2.50	2.469	53,577	36,823	29,570	25,308	22,430	20,323	17,394	15,416	10,595	8,509	7,282
3.00	3.068	94,714	65,097	52,275	44,741	39,653	35,928	30,750	27,253	18,731	15,042	12,874
3.50	3.548	138,676	95,311	76,538	65,507	58,058	52,604	45,023	39,903	27,425	22,023	18,849
4.00	4.026	193,187	132,777	106,624	91,257	80,879	73,282	62,720	55,538	38,205	30,680	26,258
5.00	5.047	349,503	240,211	192,898	165,096	146,322	132,578	113,470	100,566	69,118	55,505	47,505
6.00	6.065	565,926	388,958	312,347	267,329	236,928	214,674	183,733	162,840	111,919	89,875	76,921
8.00	7.981	1,162,762	799,160	641,754	549,258	486,797	441,074	377,502	334,573	229,950	184,658	158,043
10.00	10.020	2,111,887	1,451,488	1,165,596	997,600	884,154	801,108	685,645	607,674	417,651	335,388	287,049
12.00	11.938	3,343,383	2,297,888	1,845,285	1,579,326	1,399,727	1,268,254	1,085,462	962,025	661,194	530,962	454,435

For SI: 1 inch = 25.4 mm, 1 foot = 304.8 mm, 1 cubic foot per hour = 0.0283 m³/h, 1 pound per square inch = 6.895 kPa.

TABLE 402.3(12)
PIPE SIZING TABLE FOR 50 POUNDS PRESSURE CAPACITY OF PIPES OF DIFFERENT DIAMETERS AND LENGTHS IN CUBIC FEET PER HOUR FOR AN INITIAL PRESSURE OF 50.0 PSI WITH A 10-PERCENT PRESSURE DROP AND A GAS OF 0.6 SPECIFIC GRAVITY

PIPE SIZE OF SCHEDULE 40 STANDARD PIPE (inches)	INTERNAL DIAMETER (inches)	TOTAL EQUIVALENT LENGTH OF PIPE (feet)										
		50	100	150	200	250	300	400	500	1000	1500	2000
1.00	1.049	1,2993	8,930	7,171	6,138	5,440	4,929	4,218	3,739	2,570	2,063	1,766
1.25	1.380	26,676	18,335	14,723	12,601	11,168	10,119	8,661	7,676	5,276	4,236	3,626
1.50	1.610	39,970	27,471	22,060	18,881	16,733	15,162	12,976	11,501	7,904	6,348	5,433
2.00	2.067	76,977	52,906	42,485	36,362	32,227	29,200	24,991	22,149	15,223	12,225	10,463
2.50	2.469	122,690	84,324	67,715	57,955	51,365	46,540	39,832	35,303	24,263	19,484	16,676
3.00	3.068	216,893	149,070	119,708	102,455	90,804	82,275	70,417	62,409	42,893	34,445	29,480
3.50	3.548	317,564	218,260	175,271	150,009	132,950	120,463	103,100	91,376	62,802	50,432	43,164
4.00	4.026	442,393	304,054	244,166	208,975	185,211	167,814	143,627	127,294	87,489	70,256	60,130
5.00	5.047	800,352	550,077	441,732	378,065	335,072	303,600	259,842	230,293	158,279	127,104	108,784
6.00	6.065	1,295,955	890,703	715,266	612,175	542,559	491,598	420,744	372,898	256,291	205,810	176,147
8.00	7.981	2,662,693	1,830,054	1,469,598	1,257,785	1,114,752	1,010,046	864,469	766,163	526,579	422,862	361,915
10.00	10.020	4,836,161	3,323,866	2,669,182	2,28,4474	2,024,687	1,834,514	1,570,106	1,391,556	956,409	768,030	657,334
12.00	11.938	7,656,252	5,262,099	4,225,651	3,616,611	3,205,335	2,904,266	2,485,676	2,203,009	1,514,115	1,215,888	1,040,643

For SI: 1 inch = 25.4 mm, 1 foot = 304.8 mm, 1 cubic foot per hour = 0.0283 m³/h, 1 pound per square inch = 6.895 kPa.

TABLE 402.3(13) – TABLE 402.3(15)

GAS PIPING INSTALLATIONS

TABLE 402.3(13)
MAXIMUM CAPACITY OF SEMI-RIGID TUBING IN CUBIC FEET OF GAS PER HOUR FOR GAS
PRESSURES OF 0.5 PSI OR LESS AND A PRESSURE DROP OF 0.3-INCH WATER COLUMN
Based on a 0.60 Specific Gravity Gas

OUTSIDE DIAMETER (inch)	LENGTH OF TUBING (feet)													
	10	20	30	40	50	60	70	80	90	100	125	150	175	200
$3/8$	20	14	11	10	9	8	7	7	6	6	5	5	4	4
$1/2$	42	29	23	20	18	16	15	14	13	12	11	10	9	8
$5/8$	86	59	47	40	36	33	30	28	26	25	22	20	18	17
$3/4$	150	103	83	71	63	57	52	49	46	43	38	35	32	30
$7/8$	212	146	117	100	89	81	74	69	65	61	54	49	45	42

For SI: 1 inch = 25.4 mm, 1 foot = 304.8 mm, 1 cubic foot per hour = 0.0283 m³/h, 1 pound per square inch = 6.895 kPa, 1-inch water column = 0.2488 kPa.

TABLE 402.3(14)
MAXIMUM CAPACITY OF SEMI-RIGID TUBING IN CUBIC FEET OF GAS PER HOUR FOR
GAS PRESSURES OF 0.5 PSI OR LESS AND A PRESSURE DROP OF 0.5-INCH WATER COLUMN
Based on a 0.60 Specific Gravity Gas

OUTSIDE DIAMETER (inch)	LENGTH OF TUBING (feet)													
	10	20	30	40	50	60	70	80	90	100	125	150	175	200
$3/8$	27	18	15	13	11	10	9	9	8	8	7	6	6	5
$1/2$	56	38	31	26	23	21	19	18	17	16	14	13	12	11
$5/8$	113	78	62	53	47	43	39	37	34	33	29	26	24	22
$3/4$	197	136	109	93	83	75	69	64	60	57	50	46	42	39
$7/8$	280	193	155	132	117	106	98	91	85	81	71	65	60	55

For SI: 1 inch = 25.4 mm, 1 foot = 304.8 mm, 1 cubic foot per hour = 0.0283 m³/h, 1 pound per square inch = 6.895 kPa, 1-inch water column = 0.2488 kPa.

TABLE 402.3(15)
MAXIMUM CAPACITY OF SEMI-RIGID TUBING IN CUBIC FEET OF GAS PER HOUR FOR GAS PRESSURES
OF 0.5 PSI OR LESS AND A PRESSURE DROP OF 1.0-INCH WATER COLUMN
Based on a 0.60 Specific Gravity Gas

USE THIS TABLE TO SIZE TUBING FROM HOUSE LINE REGULATOR TO THE APPLIANCE. DIAMETER: INSIDE (OUTSIDE)							
LENGTH (feet)	1/4 inch (0.315 in.)	3/8 inch (0.430 in.)	1/2 inch (0.545 in.)	5/8 inch (0.666 in.)	3/4 inch (0.785 in.)	1 inch (1.025 in.)	1 1/4 inch (1.265 in.)
10	42	95	177	300	461	928	1,612
15	34	76	142	241	370	745	1,294
20	29	65	122	206	317	638	1,108
30	23	52	98	165	255	512	890
40	20	45	84	142	218	439	761
50	18	40	74	125	193	389	675
60	16	36	67	114	175	352	611
70	15	33	62	105	161	324	563
80	14	31	57	97	150	301	523
90	13	29	54	91	140	283	491
100	12	27	51	86	133	267	464
125	11	24	45	76	118	237	411
150	10	22	41	69	107	215	372
175	9	20	38	64	98	197	343
200	8	19	35	59	91	184	319
250	7	17	31	53	81	163	283
300	7	15	28	48	73	147	256

For SI: 1 inch = 25.4 mm, 1 foot = 304.8 mm, 1 cubic foot per hour = 0.0283 m³/h, 1 pound per square inch = 6.895 kPa,
1-inch water column = 0.2488 kPa.

TABLE 402.3(16)
MAXIMUM CAPACITY OF SEMI-RIGID TUBING IN CUBIC FEET PER HOUR FOR A
GAS PRESSURE OF 2 PSI OR LESS AND A PRESSURE DROP OF 17-INCH WATER COLUMN
Based on a 0.60 Specific Gravity Gas

	DIAMETER: INSIDE (OUTSIDE)						
LENGTH (feet)	1/4 inch (0.315 in.)	3/8 inch (0.430 in.)	1/2 inch (0.545 in.)	5/8 inch (0.666 in.)	3/4 inch (0.785 in.)	1 inch (1.025 in.)	1 1/4 inch (1.265 in.)
10	201	454	845	1,435	2,200	4,428	7,690
15	161	364	678	1,152	1,766	3,556	6,175
20	138	312	581	986	1,512	3,044	5,285
30	111	250	466	792	1,214	2,444	4,244
40	95	214	399	678	1,039	2,092	3,632
50	84	190	354	601	921	1,854	3,219
60	76	172	320	544	834	1,680	2,917
70	70	158	295	501	768	1,545	2,684
80	65	147	274	466	714	1,438	2,496
90	61	139	257	437	670	1,349	2,342
100	58	131	243	413	633	1,274	2,213
125	51	116	215	366	561	1,129	1,961
150	46	105	195	332	508	1,023	1,777
175	43	96	180	305	468	941	1,635
200	40	90	167	284	435	876	1,521
250	35	80	148	251	386	776	1,348
300	32	72	134	228	349	703	1,121

For SI: 1 inch = 25.4 mm, 1 foot = 304.8 mm, 1 cubic foot per hour = 0.0283 m³/h, 1 pound per square inch = 6.895 kPa, 1-inch water column = 0.2488 kPa.

TABLE 402.3(17)
MAXIMUM CAPACITY OF SEMI-RIGID TUBING IN CUBIC FEET OF GAS PER HOUR FOR A
GAS PRESSURE OF 2.0 PSI OR LESS AND A PRESSURE DRIP OF 1.0 PSI
Based on a 0.60 Specific Gravity Gas

NOMINAL TUBING DIAMETER INSIDE (inches)	INTERNAL DIAMETER (inches)	LENGTH OF TUBING (feet)																	
		5	10	15	20	30	40	50	60	70	80	90	100	125	150	175	200	250	300
1/4	0.315	459	306	242	204	163	139	122	110	100	93	87	82	72	65	59	54	49	43
3/8	0.430	1,071	722	569	484	382	323	285	255	234	217	204	191	168	151	139	124	119	102
1/2	0.545	2,040	1,385	1,088	918	731	620	548	493	450	416	391	365	323	289	268	246	217	195
5/8	0.666	3,527	2,363	1,827	1,581	1,258	1,062	935	850	773	722	671	629	552	497	459	425	374	336
3/4	0.785	5,524	3,697	2,932	2,507	1,955	1,700	1,487	1,326	1,215	1,130	1,045	986	871	782	718	671	586	527
1	1.025	8,923	6,459	5,269	4,589	3,739	3,229	2,847	2,592	2,380	2,252	2,125	1,997	1,785	1,615	1,530	1,445	1,275	1,147
1 1/4	1.265	17,847	12,748	10,198	8,923	7,309	6,374	5,694	5,184	4,759	4,419	4,164	3,994	3,627	3,229	3,017	2,804	2,507	2,295
1 1/2	1.505	26,345	18,696	15,297	12,748	11,048	9,348	8,328	7,649	6,969	6,544	6,119	5,779	5,184	4,759	4,419	4,164	3,654	3,399
2	1.985	49,291	34,843	28,894	24,645	20,396	16,997	15,297	14,447	12,748	11,898	11,473	10,623	9,603	8,838	8,243	7,649	6,884	6,289
2 1/2	2.465	76,485	54,390	44,192	38,243	30,594	27,195	23,795	22,096	20,396	18,696	17,847	16,997	15,297	13,597	13,172	11,898	10,623	9,773

For SI: 1 inch = 25.4 mm, 1 foot = 304.8 mm, 1 cubic foot per hour = 0.0283 m³/h. 1 pound per square inch = 6.895 kPa.

TABLE 402.3(18)
MAXIMUM CAPACITY OF SEMI-RIGID TUBING IN CUBIC FEET OF GAS PER HOUR FOR
A GAS PRESSURE OF 5.0 PSI OR LESS AND A PRESSURE DROP OF 3.5 PSI
Based on a 0.60 Specific Gravity Gas

NOMINAL TUBING DIAMETER INSIDE (inches)	INTERNAL DIAMETER (inches)	LENGTH OF TUBING (feet)																	
		5	10	15	20	30	40	50	60	70	80	90	100	125	150	175	200	250	300
1/4	0.315	791	527	417	351	281	239	209	190	173	161	149	141	124	111	101	94	85	75
3/8	0.430	1,845	1,245	981	835	659	556	490	439	403	373	351	329	290	261	240	214	205	176
1/2	0.545	3,514	2,387	1,874	1,581	1,259	1,069	944	849	776	717	674	630	556	498	461	425	373	337
5/8	0.666	6,076	4,070	3,148	2,723	2,167	1,830	1,611	1,464	1,332	1,245	1,157	1,083	952	857	791	732	644	578
3/4	0.785	9,517	6,369	5,051	4,319	3,368	2,928	2,562	2,284	2,094	1,947	1,801	1,698	1,501	1,347	1,237	1,142	1,010	906
1	1.025	15,374	11,127	9,078	7,906	6,442	5,564	4,905	4,466	4,100	3,880	3,660	3,441	3,075	2,782	2,635	2,489	2,196	1,977
1 1/4	1.265	30,747	21,962	17,570	15,374	12,592	10,981	9,810	8,931	8,199	7,614	7,174	6,881	6,076	5,564	5,198	4,832	4,319	3,853
1 1/2	1.505	45,388	32,211	26,355	21,962	19,034	16,106	14,349	13,177	12,006	11,274	10,542	9,956	8,931	8,199	7,614	7,174	6,296	5,857
2	1.985	84,920	60,030	49,781	42,460	35,139	29,283	26,356	24,890	21,962	20,498	19,766	18,302	16,545	15,227	14,202	13,177	11,860	10,835
2 1/2	2.465	131,773	93,705	76,135	65,886	52,709	46,853	40,996	38,068	35,139	32,211	30,747	29,283	26,355	23,426	22,694	20,498	18,302	16,838

For SI: 1 inch = 25.4 mm, 1 foot = 304.8 mm, 1 cubic foot per hour = 0.0283 m³/h, 1 pound per square inch = 6.895 kPa.

TABLE 402.3(19)
MAXIMUM CAPACITY OF CSST IN CUBIC FEET PER HOUR FOR GAS PRESSURE OF
0.5 PSI OR LESS AND A PRESSURE DROP OF 0.5-INCH WATER COLUMN[a]
Based on a 0.60 Specific Gravity Gas

EHD[b] FLOW DESIGNATION	TUBING LENGTH (feet)																
	5	10	15	20	25	30	40	50	60	70	80	90	100	150	200	250	300
13	46	32	25	22	19	18	15	13	12	11	10	10	9	7	6	5	5
15	63	44	35	31	27	25	21	19	17	16	15	14	13	10	9	8	7
18	115	82	66	58	52	47	41	37	34	31	29	28	26	20	18	16	15
19	134	95	77	67	60	55	47	42	38	36	33	32	30	23	21	19	17
23	225	161	132	116	104	96	83	75	68	63	60	57	54	42	38	34	32
25	270	192	157	137	122	112	97	87	80	74	69	65	62	48	44	39	36
30	471	330	267	231	206	188	162	144	131	121	113	107	101	78	71	63	57
31	546	383	310	269	240	218	188	168	153	141	132	125	118	91	82	74	67

For SI: 1 foot = 304.8 mm, 1 cubic foot per hour = 0.0283 m³/h, 1 pound per square inch = 6.895 kPa,
1-inch water column = 0.2488 kPa, 1 degree = 0.01745 rad.

a. Table includes losses for four 90-degree bends and two end fittings. Tubing runs with larger numbers of bends and/or fittings shall be increased by an equivalent length of tubing to the following equation: $L = 1.3n$ where L is additional length (feet) of tubing and n is the number of additional fittings and/or bends.

b. EHD — Equivalent Hydraulic Diameter — A measure of the relative hydraulic efficiency between different tubing sizes. The greater the value of EHD, the greater the gas capacity of the tubing.

TABLE 402.3(20)
MAXIMUM CAPACITY OF CSST IN CUBIC FEET PER HOUR FOR A GAS PRESSURE
OF 0.5 PSI OR LESS AND A PRESSURE DROP OF 3-INCH WATER COLUMN[a]
Based on a 0.60 Specific Gravity Gas

EHD[b] FLOW DESIGNATION	TUBING LENGTH (feet)																
	5	10	15	20	25	30	40	50	60	70	80	90	100	150	200	250	300
13	120	83	67	57	51	46	39	35	32	29	27	26	24	19	17	15	13
15	160	112	90	78	69	63	54	48	44	41	38	36	34	27	23	21	19
18	277	197	161	140	125	115	100	89	82	76	71	67	63	52	45	40	37
19	327	231	189	164	147	134	116	104	95	88	82	77	73	60	52	46	42
23	529	380	313	273	245	225	196	176	161	150	141	133	126	104	91	82	75
25	649	462	379	329	295	270	234	210	192	178	167	157	149	122	106	95	87
30	1,182	828	673	580	518	471	407	363	330	306	285	268	254	206	178	159	144
31	1,365	958	778	672	599	546	471	421	383	355	331	311	295	240	207	184	168

For SI: 1 inch = 25.4 mm, 1 foot = 304.8 mm, 1 cubic foot per hour = 0.0283 m³/h, 1 pound per square inch = 6.895 kPa.

a. Table includes losses for four 90-degree bends and two end fittings. Tubing runs with larger numbers of bends and/or fittings shall be increased by an equivalent length of tubing to the following equation: $L = 1.3n$ where L is additional length (feet) of tubing and n is the number of additional fittings and/or bends.

b. EHD — Equivalent Hydraulic Diameter — A measure of the relative hydraulic efficiency between different tubing sizes. The greater the value of EHD, the greater the gas capacity of the tubing.

TABLE 402.3(21)
MAXIMUM CAPACITY OF CSST IN CUBIC FEET PER HOUR FOR A GAS PRESSURE OF
0.5 PSI OR LESS AND A PRESSURE DROP OF 6-INCH WATER COLUMN[a]
Based on a 0.60 Specific Gravity Gas

EHD[b] FLOW DESIGNATION	TUBING LENGTH (feet)																
	5	10	15	20	25	30	40	50	60	70	80	90	100	150	200	250	300
13	173	120	96	83	74	67	57	51	46	42	39	37	35	28	24	21	19
15	229	160	130	112	99	90	78	69	63	58	54	51	48	39	34	30	27
18	389	277	227	197	176	161	140	125	115	106	100	94	89	73	63	57	52
19	461	327	267	231	207	189	164	147	134	124	116	109	104	85	73	66	60
23	737	529	436	380	342	313	273	245	225	209	196	185	176	145	126	114	104
25	911	649	532	462	414	379	329	295	270	250	234	221	210	172	149	134	122
30	1,687	1,182	960	828	739	673	580	518	471	435	407	383	363	294	254	226	206
31	1,946	1,365	1,110	958	855	778	672	599	546	505	471	444	421	342	295	263	240

For SI: 1 inch = 25.4 mm, 1 foot = 304.8 mm, 1 cubic foot per hour = 0.0283 m³/h, 1 pound per square inch = 6.895 kPa.

a. Table includes losses for four 90-degree bends and two end fittings. Tubing runs with larger numbers of bends and/or fittings shall be increased by an equivalent length of tubing to the following equation: $L = 1.3n$ where L is additional length (feet) of tubing and n is the number of additional fittings and/or bends.

b. EHD — Equivalent Hydraulic Diameter — A measure of the relative hydraulic efficiency between different tubing sizes. The greater the value of EHD, the greater the gas capacity of the tubing.

TABLE 402.3(22) – TABLE 402.3(23)

GAS PIPING INSTALLATIONS

TABLE 402.3(22)
MAXIMUM CAPACITY OF CSST IN CUBIC FEET PER HOUR FOR A GAS PRESSURE
OF 2 PSI AND A PRESSURE DROP OF 1PSI[a]
Based on a 0.60 Specific Gravity Gas

EHD[b] FLOW DESIGNATION	TUBING LENGTH (feet)													
	10	25	30	40	50	75	80	110	150	200	250	300	400	500
13	270	166	151	129	115	93	89	79	64	55	49	44	38	34
15	353	220	200	172	154	124	120	107	87	75	67	61	52	46
18	587	374	342	297	266	218	211	189	155	135	121	110	96	86
19	700	444	405	351	314	257	249	222	182	157	141	129	111	100
23	1,098	709	650	567	510	420	407	366	302	263	236	217	189	170
25	1,372	876	801	696	624	512	496	445	364	317	284	260	225	202
30	2,592	1,620	1,475	1,273	1,135	922	892	795	646	557	497	453	390	348
31	2,986	1,869	1,703	1,470	1,311	1,066	1,031	920	748	645	576	525	453	404

Table does not include effect of pressure drop across line regulator. If regulator loss exceeds 3/4 psi, **DO NOT USE THIS TABLE.** Consult with regulator manufacturer for pressure drops and capacity factors. Pressure drop across regulator may vary with the flow rate.

CAUTION: Capacities shown in table may exceed maximum capacity of selected regulator. Consult with tubing manufacturer for guidance.

For SI: 1 inch = 25.4 mm, 1 foot = 304.8 mm, 1 cubic foot per hour = 0.0283 m³/h, 1 pound per square inch = 6.895 kPa.

a. Table includes losses for four 90-degree bends and two end fittings. Tubing runs with larger numbers of bends and/or fittings shall be increased by an equivalent length of tubing to the following equation: $L = 1.3n$ where L is additional length (feet) of tubing and n is the number of additional fittings and/or bends.
b. EHD — Equivalent Hydraulic Diameter — A measure of the relative hydraulic efficiency between different tubing sizes. The greater the value of EHD, the greater the gas capacity of the tubing.

TABLE 402.3(23)
MAXIMUM CAPACITY OF CSST IN CUBIC FEET PER HOUR FOR A GAS PRESSURE
OF 5 PSI AND A PRESSURE DROP OF 3.5 PSI[a]
Based on a 0.60 Specific Gravity Gas

EHD[b] FLOW DESIGNATION	TUBING LENGTH (feet)													
	10	25	30	40	50	75	80	110	150	200	250	300	400	500
13	523	322	292	251	223	180	174	154	124	107	95	86	74	66
15	674	420	382	329	293	238	230	205	166	143	128	116	100	89
18	1,084	691	632	549	492	403	391	350	287	249	223	204	177	159
19	1,304	827	755	654	586	479	463	415	339	294	263	240	208	186
23	1,995	1,289	1,181	1,031	926	763	740	665	548	478	430	394	343	309
25	2,530	1,616	1,478	1,284	1,151	944	915	820	672	584	524	479	416	373
30	4,923	3,077	2,803	2,418	2,157	1,752	1,694	1,511	1,228	1,060	945	860	742	662
31	5,959	3,543	3,228	2,786	2,486	2,021	1,955	1,744	1,418	1,224	1,092	995	858	766

Table does not include effect of pressure drop across line regulator. If regulator loss exceeds 1 psi, **DO NOT USE THIS TABLE.** Consult with regulator manufacturer for pressure drops and capacity factors. Pressure drop across regulator may vary with the flow rate.

CAUTION: Capacities shown in table may exceed maximum capacity of selected regulator. Consult with tubing manufacturer for guidance.

For SI: 1 inch = 25.4 mm, 1 foot = 304.8 mm, 1 cubic foot per hour = 0.0283 m³/h, 1 pound per square inch = 6.895 kPa.

a. Table includes losses for four 90-degree bends and two end fittings. Tubing runs with larger numbers of bends and/or fittings shall be increased by an equivalent length of tubing to the following equation: $L = 1.3n$ where L is additional length (feet) of tubing and n is the number of additional fittings and/or bends.
b. EHD — Equivalent Hydraulic Diameter — A measure of the relative hydraulic efficiency between different tubing sizes. The greater the value of EHD, the greater the gas capacity of the tubing.

TABLE 402.3(24)
MULTIPLIERS TO BE USED WITH TABLES 402(1) THROUGH 402(12)
WHEN THE SPECIFIC GRAVITY OF THE GAS IS OTHER THAN 0.60

SPECIFIC GRAVITY	MULTIPLIER	SPECIFIC GRAVITY	MULTIPLIER
.35	1.31	1.00	.78
.40	1.23	1.10	.74
.45	1.16	1.20	.71
.50	1.10	1.30	.68
.55	1.04	1.40	.66
.60	1.00	1.50	.63
.65	.96	1.60	.61
.70	.93	1.70	.59
.75	.90	1.80	.58
.80	.87	1.90	.56
.85	.84	2.00	.55
.90	.82	2.10	.54

TABLE 402.3(25)
PIPE SIZING
Sizing between First Stage (High-Pressure Regulator) and
Second Stage (Low-Pressure Regulator)

Maximum undiluted propane capacities listed are based on a 10 psi first-stage setting and 1 psi pressure drop. Capacities in thousands of Btu per hour

PIPE LENGTH (feet)	SCHEDULE 40 PIPE SIZE 1 psi drop								
	1/2" 0.622	3/4" 0.824	1" 1.049	1 1/4" 1.38	1 1/2" 1.61	2" 2.067	3" 3.068	3 1/2" 3.548	4" 4.026
30	1,843	3,854	7,259	14,904	22,331	43,008	121,180	177,425	247,168
40	1,577	3,298	6,213	12,756	19,113	36,809	103,714	151,853	211,544
50	1,398	2,923	5,507	11,306	16,939	32,623	91,920	134,585	187,487
60	1,267	2,649	4,989	10,244	15,348	29,559	83,286	121,943	169,877
70	1,165	2,437	4,590	9,424	14,120	27,194	76,622	112,186	156,285
80	1,084	2,267	4,270	8,767	13,136	25,299	71,282	104,368	145,393
90	1,017	2,127	4,007	8,226	12,325	23,737	66,882	97,925	136,417
100	961	2,009	3,785	7,770	11,642	22,422	63,176	92,499	128,859
150	772	1,613	3,039	6,240	9,349	18,005	50,733	74,280	103,478
200	660	1,381	2,601	5,340	8,002	15,410	43,421	63,574	88,564
250	585	1,224	2,305	4,733	7,092	13,658	38,483	56,345	78,493
300	530	1,109	2,089	4,289	6,426	12,375	34,868	51,052	71,120
350	488	1,020	1,922	3,945	5,911	11,385	32,078	46,967	65,430
400	454	949	1,788	3,670	5,499	10,591	29,843	43,694	60,870
450	426	890	1,677	3,444	5,160	9,938	28,000	40,997	57,112
500	402	841	1,584	3,253	4,874	9,387	26,449	38,725	53,948
600	364	762	1,436	2,948	4,416	8,505	23,965	35,088	48,880
700	335	701	1,321	2,712	4,063	7,825	22,047	32,280	44,969
800	312	652	1,229	2,523	3,780	7,279	20,511	30,031	41,835
900	293	612	1,153	2,367	3,546	6,830	19,245	28,177	39,253
1,000	276	578	1,089	2,236	3,350	6,452	18,178	26,616	37,078
1,500	222	464	875	1,795	2,690	5,181	14,598	21,373	29,775
2,000	190	397	748	1,537	2,302	4,434	12,494	18,293	25,483

For SI: 1 inch = 25.4 mm, 1 foot = 304.8 mm, 1 cubic foot per hour = 0.0283 m³/h, 1 pound per square inch = 6.895 kPa.

TABLE 402.3(26) – TABLE 402.3(27)   GAS PIPING INSTALLATIONS

TABLE 402.3(26)
PIPE SIZING
Sizing between Single or Second Stage (Low-Pressure Regulator) and Appliance

Maximum undiluted propane capacities listed are based on 11-inch water column
setting and a 0.5-inch water column pressure drop. Capacities in thousands of Btu per hour

PIPE LENGTH (feet)	NOMINAL PIPE SIZE, SCHEDULE 40								
	1/2" 0.622	3/4" 0.824	1" 1.049	1 1/4" 1.38	1 1/2" 1.61	2" 2.067	3" 3.068	3 1/2" 3.548	4" 4.026
10	291	608	1,146	2,353	3,525	6,789	19,130	28,008	39,018
20	200	418	788	1,617	2,423	4,666	13,148	19,250	26,817
30	161	336	632	1,299	1,946	3,747	10,558	15,458	21,535
40	137	287	541	1,111	1,665	3,207	9,036	13,230	18,431
50	122	255	480	985	1,476	2,842	8,009	11,726	16,335
60	110	231	435	892	1,337	2,575	7,256	10,625	14,801
80	94	198	372	764	1,144	2,204	6,211	9,093	12,668
100	84	175	330	677	1,014	1,954	5,504	8,059	11,227
125	74	155	292	600	899	1,731	4,878	7,143	9,950
150	67	141	265	544	815	1,569	4,420	6,472	9,016
200	58	120	227	465	697	1,343	3,783	5,539	7,716
250	51	107	201	412	618	1,190	3,353	4,909	6,839
300	46	97	182	374	560	1,078	3,038	4,448	6,196
350	43	89	167	344	515	992	2,795	4,092	5,701
400	40	83	156	320	479	923	2,600	3,807	5,303

For SI: 1 inch = 25.4 mm, 1 foot = 304.8 mm, 1 cubic foot per hour = 0.0283 m³/h, 1 pound per square inch = 6.895 kPa.

TABLE 402.3(27)
COPPER TUBE SIZING SIZING BETWEEN FIRST STAGE (HIGH-PRESSURE REGULATOR)
AND SECOND STAGE (LOW-PRESSURE REGULATOR)

Maximum undiluted propane capacities listed are based on a
10 psi first stage setting and 1 psi drop. Capacities in thousands of Btu per hour

TUBING LENGTH (feet)	OUTSIDE DIAMETER COPPER TUBING, TYPE L				
	3/8" 0.315	1/2" 0.430	5/8" 0.545	3/4" 0.666	7/8" 0.785
30	309	700	1,303	2,205	3,394
40	265	599	1,115	1,887	2,904
50	235	531	988	1,672	2,574
60	213	481	896	1,515	2,332
70	196	443	824	1,394	2,146
80	182	412	767	1,297	1,996
90	171	386	719	1,217	1,873
100	161	365	679	1,149	1,769
150	130	293	546	923	1,421
200	111	251	467	790	1,216
250	90	222	414	700	1,078
300	89	201	375	634	976
350	82	185	345	584	898
400	76	172	321	543	836
450	71	162	301	509	784
500	68	153	284	481	741
600	61	138	258	436	671
700	56	127	237	401	617
800	52	118	221	373	574
900	49	111	207	350	539
1,000	46	105	195	331	509
1,500	37	84	157	266	409
2,000	32	72	134	227	350

For SI: 1 inch = 25.4 mm, 1 foot = 304.8 mm, 1 cubic foot per hour = 0.0283 m³/h,
1 pound per square inch = 6.895 kPa.

TABLE 402.3(28)
COPPER TUBE SIZING SIZING BETWEEN SINGLE OR SECOND STAGE
(LOW-PRESSURE REGULATOR) AND APPLIANCE

Maximum undiluted propane capacities are based on 11-inch water column
setting and a 0.5-inch water column pressure drop. Capacities in thousands of Btu per hour

TUBING LENGTH (feet)	OUTSIDE DIAMETER COPPER TUBING, TYPE L				
	3/8" 0.315	1/2" 0.430	5/8" 0.545	3/4" 0.666	7/8" 0.785
10	49	110	206	348	536
20	34	76	141	239	368
30	27	61	114	192	296
40	23	52	97	164	253
50	20	46	86	146	224
60	19	42	78	132	203
80	16	36	67	113	174
100	14	32	59	100	154
125	12	28	52	89	137
150	11	26	48	80	124
200	10	22	41	69	106
250	9	19	36	61	94
300	8	18	33	55	85
350	7	16	30	51	78
400	7	15	28	47	73

For SI: 1 inch = 25.4 mm, 1 foot = 304.8 mm, 1 cubic foot per hour = 0.0283 m³/h,
           1 pound per square inch = 6.895 kPa.

TABLE 402.3(29)
MAXIMUM CAPACITY OF CSST IN THOUSANDS OF BTU PER HOUR OF
UNDILUTED LIQUEFIED PETROLEUM GASES AT A PRESSURE OF 11-INCH WATER
COLUMN AND A PRESSURE DROP OF 0.5-INCH WATER COLUMN[a]
Based on a 0.60 Specific Gravity Gas

EHD[b] FLOW DESIGNATION	TUBING LENGTH (feet)																
	5	10	15	20	25	30	40	50	60	70	80	90	100	150	200	250	300
13	72	50	39	34	30	28	23	20	19	17	15	15	14	11	9	8	8
15	99	69	55	49	42	39	33	30	26	25	23	22	20	15	14	12	11
18	181	129	104	91	82	74	64	58	53	49	45	44	41	31	28	25	23
19	211	150	121	106	94	87	74	66	60	57	52	50	47	36	33	30	26
23	355	254	208	183	164	151	131	118	107	99	94	90	85	66	60	53	50
25	426	303	248	216	192	177	153	137	126	117	109	102	98	75	69	61	57
30	744	521	422	365	325	297	256	227	207	191	178	169	159	123	112	99	90
31	863	605	490	425	379	344	297	265	241	222	208	197	186	143	129	117	107

For SI: 1 inch = 25.4 mm, 1 foot = 304.8 mm, 1 cubic foot per hour = 0.0283 m³/h, 1 pound per square inch = 6.895 kPa.

a. Table includes losses for four 90-degree bends and two end fittings. Tubing runs with larger numbers of bends and/or fittings shall be increased by an equivalent length of tubing to the following equation: $L = 1.3n$ where L is additional length (feet) of tubing and n is the number of additional fittings and/or bends.

b. EHD — Equivalent Hydraulic Diameter — A measure of the relative hydraulic efficiency between different tubing sizes. The greater the value of EHD, the greater the gas capacity of the tubing.

TABLE 402.3(30) – TABLE 402.3(31)   GAS PIPING INSTALLATIONS

TABLE 402.3(30)
MAXIMUM CAPACITY OF CSST IN THOUSANDS OF BTU PER HOUR OF UNDILUTED
LIQUEFIED PETROLEUM GASES AT A PRESSURE OF 2 PSI AND A PRESSURE DROP OF 1 PSI[a]

Based on 1.52 Specific Gravity Gas

EHD[b] FLOW DESIGNATION	TUBING LENGTH (feet)													
	10	25	30	40	50	75	80	110	150	200	250	300	400	500
13	426	262	238	203	181	147	140	124	101	86	77	69	60	53
15	558	347	316	271	243	196	189	169	137	118	105	96	82	72
18	927	591	540	469	420	344	333	298	245	213	191	173	151	135
19	1,106	701	640	554	496	406	393	350	287	248	222	203	175	158
23	1,735	1,120	1,027	896	806	663	643	578	477	415	373	343	298	268
25	2,168	1,384	1,266	1,100	986	809	768	703	575	501	448	411	355	319
30	4,097	2,560	2,331	2,012	1,794	1,457	1,410	1,256	1,021	880	785	716	616	550
31	4,720	2,954	2,692	2,323	2,072	1,685	1,629	1,454	1,182	1,019	910	829	716	638

Table does not include effect of pressure drop across the line regulator. If regulator loss exceeds $1/2$ psi (based on 13-inch water column outlet pressure), **DO NOT USE THIS TABLE**. Consult with regulator manufacturer for pressure drops and capacity factors. Pressure drops across a regulator may vary with flow rate.

CAUTION: Capacities shown in table may exceed maximum capacity for a selected regulator. Consult with regulator or tubing manufacturer for guidance.

For SI: 1 inch = 25.4 mm, 1 foot = 304.8 mm, 1 cubic foot per hour = 0.0283 m³/h, 1 pound per square inch = 6.895 kPa.

a. Table includes losses for four 90-degree bends and two end fittings. Tubing runs with larger number of bends and/or fittings shall be increased by an equivalent length of tubing according to the following equation: $L-1.3n$ where L is additional length (feet) of tubing and n is the number of additional fittings and/or bends.

b. EHD — Equivalent Hydraulic Diameter — A measure of the relative hydraulic efficiency between different tubing sizes. The greater the value of EHD, the greater the gas capacity of the tubing.

TABLE 402.3(31)
MAXIMUM CAPACITY OF CSST IN THOUSANDS OF BTU PER HOUR OF UNDILUTED LIQUEFIED PETROLEUM
GASES AT A PRESSURE OF 5 PSI AND A PRESSURE DROP OF 3.5 PSI[a]

Based on a 1.52 specific Gravity Gas

EHD[b] FLOW DESIGNATION	TUBING LENGTH (feet)													
	10	25	30	40	50	75	80	100	150	200	250	300	400	500
13	826	509	461	396	352	284	275	243	196	169	150	136	117	104
15	1,065	664	603	520	463	376	363	324	262	226	202	183	156	140
18	1,713	1,092	999	867	777	637	618	553	453	393	352	322	279	251
19	2,061	1,307	1,193	1,033	926	757	731	656	535	464	415	379	328	294
23	3,153	2,037	1,866	1,629	1,463	1,206	1,169	1,051	866	755	679	622	542	488
25	3,999	2,554	2,336	2,029	1,819	1,492	1,446	1,296	1,062	923	828	757	657	589
30	7,829	4,864	4,430	3,822	3,409	2,769	2,677	2,388	1,941	1,675	1,493	1,359	1,173	1,046
31	8,945	5,600	5,102	4,404	3,929	3,194	3,090	2,756	2,241	1,934	1,726	1,572	1,356	1,210

Table does not include effect of pressure drop across line regulator. If regulator loss exceeds 1 psi, **DO NOT USE THIS TABLE**. Consult with regulator manufacturer for pressure drops and capacity factors. Pressure drop across regulator may vary with the flow rate. **CAUTION:** Capacities shown in table may exceed maximum capacity of selected regulator. Consult with tubing manufacturer for guidance.

For SI: 1 inch = 25.4 mm, 1 foot = 304.8 mm, 1 cubic foot per hour = 0.0283 m³/h, 1 pound per square inch = 6.895 kPa.

a. Table includes losses for four 90-degree bends and two end fittings. Tubing runs with larger numbers of bends and/or fittings shall be increased by an equivalent length of tubing to the following equation: $L = 1.3n$ where L is additional length (feet) of tubing and n is the number of additional fittings and/or bends.

b. EHD — Equivalent Hydraulic Diameter — A measure of the relative hydraulic efficiency between different tubing sizes. The greater the value of EHD, the greater the gas capacity of the tubing.

TABLE 402.3(32)
POLYETHYLENE PLASTIC PIPE SIZING
SIZING BETWEEN FIRST-STAGE AND SECOND-STAGE REGULATOR

Maximum undiluted propane capacities listed are based on 10 psi
first-stage setting and 1 psi pressure drop. Capacities in thousands of Btu per hour

PLASTIC PIPE LENGTH (feet)	PLASTIC PIPE NOMINAL OUTSIDE DIAMETER (IPS) (dimensions in parenthesis are inside diameter)					
	1/2 in. SDR 9.33 (.660)	3/4 in. SDR 11.0 (.860)	1 in. SDR 11.00 (1.077)	1-1/4 in. SDR 10.00 (1.328)	1-1/2 in. SDR 11.00 (1.554)	2 in. SDR 11.00 (1.943)
30	2,143	4,292	7,744	13,416	20,260	36,402
40	1,835	3,673	6,628	11,482	17,340	31,155
50	1,626	3,256	5,874	10,176	15,368	27,612
60	1,473	2,950	5,322	9,220	13,924	25,019
70	1,355	2,714	4,896	8,483	12,810	23,017
80	1,261	2,525	4,555	7,891	11,918	21,413
90	1,183	2,369	4,274	7,404	11,182	20,091
100	1,117	2,238	4,037	6,994	10,562	18,978
125	990	1,983	3,578	6,199	9,361	16,820
150	897	1,797	3,242	5,616	8,482	15,240
175	826	1,653	2,983	5,167	7,803	14,020
200	678	1,539	2,775	4,807	7,259	13,043
225	721	1,443	2,603	4,510	6,811	12,238
250	681	1,363	2,459	4,260	6,434	11,560
275	646	1,294	2,336	4,046	6,111	10,979
300	617	1,235	2,228	3,860	5,830	10,474
350	567	1,136	2,050	3,551	5,363	9,636
400	528	1,057	1,907	3,304	4,989	8,965
450	495	992	1,789	3,100	4,681	8,411
500	468	937	1,690	2,928	4,422	7,945
600	424	849	1,531	2,653	4,007	7,199
700	390	781	1,409	2,441	3,686	6,623
800	363	726	1,311	2,271	3,429	6,161
900	340	682	1,230	2,131	3,217	5,781
1,000	322	644	1,162	2,012	3,039	5,461
1,500	258	517	933	1,616	2,441	4,385
2,000	221	443	798	1,383	2,089	3,753

For SI: 1 inch = 25.4 mm, 1 foot = 304.8 mm, 1 cubic foot per hour = 0.0283 m³/h, 1 pound per square inch = 6.895 kPa.

TABLE 402.3(33)   GAS PIPING INSTALLATIONS

TABLE 402.3(33)
POLYETHYLENE PLASTIC TUBE SIZING

Sizing between Single or Second-Stage Regulator and Building Maximum
undiluted propane capacities listed are based on 10 psi first-stage
setting and 1 psi pressure drop. Capacities in thousands of Btu per hour

PLASTIC TUBING LENGTH (feet)	PLASTIC TUBING SIZE (CTS) (dimensions in parenthesis are inside diameter)	
	1/2 in. CTS SDR 7.00 (.445)	1 in. CTS SDR 11.00 (.927)
30	762	5,225
40	653	4,472
50	578	3,964
60	524	3,591
70	482	3,304
80	448	3,074
90	421	2,884
100	397	2,724
125	352	2,414
150	319	2,188
175	294	2,013
200	273	1,872
225	256	1,757
250	242	1,659
275	230	1,576
300	219	1,503
350	202	1,383
400	188	1,287
450	176	1,207
500	166	1,140
600	151	1,033
700	139	951
800	129	884
900	121	830
1,000	114	784
1,500	92	629
2,000	79	539

For SI: 1 inch = 25.4 mm, 1 foot = 304.8 mm, 1 cubic foot per hour = 0.0283 m³/h,
1 pound per square inch = 6.895 kPa.

TABLE 402.3(34)
POLYETHYLENE PLASTIC TUBE SIZING

Sizing between Single or Second-Stage Regulator and Building Maximum undiluted
propane capacities listed are based on 11-inch water column setting and
a 0.5 inch water column pressure drop. Capacities in 1000 Btu per hour

PLASTIC TUBING LENGTH (feet)	PLASTIC TUBING SIZE (CTS) (dimensions in parenthesis are inside diameter)	
	1/2 in. CTS SDR 7.00 (.445)	1 in. CTS SDR 11.00 (.927)
10	121	829
20	83	569
30	67	457
40	57	391
50	51	347
60	46	314
70	42	289
80	39	269
90	37	252
100	35	238
125	31	211
150	28	191
175	26	176
200	24	164
225	22	154
250	21	145
275	20	138
300	19	132
350	18	121
400	16	113

For SI: 1 inch = 25.4 mm, 1 foot = 304.8 mm, 1 cubic foot per hour = 0.0283 m^3/h,
1 pound per square inch = 6.895 kPa.

SECTION 403
PIPING MATERIALS

403.1 Material application. Materials and components conforming to standards or specifications listed herein shall be permitted to be used for the appropriate applications, as prescribed and limited by this code.

403.2 Used materials. Pipe, fittings, valves, or other materials shall not be used again unless they are free of foreign materials and have been ascertained to be adequate for the service intended.

403.3 Other materials. Material not covered by the standards specifications listed herein shall be investigated and tested to determine that it is safe and suitable for the proposed service and, in addition, shall be recommended for that service by the manufacturer.

403.4 Metallic pipe. Metallic pipe shall comply with Sections 403.4.1 through 403.4.4.

403.4.1 Cast iron. Cast-iron pipe shall not be used.

403.4.2 Steel. Steel and wrought-iron pipe shall be at least of standard weight (Schedule 40) and shall comply with one of the following standards:
1. ASME B36.10, 10M
2. ASTM A 53; or
3. ASTM A 106.

403.4.3 Copper and brass. Copper and brass pipe shall not be used if the gas contains more than an average of 0.3 grains of hydrogen sulfide per 100 standard cubic feet of gas (0.7 milligrams per 100 liters). Threaded copper, brass, or aluminum alloy pipe in iron pipe sizes shall be permitted to be used with gases not corrosive to such material.

403.4.4 Aluminum. Aluminum alloy pipe shall comply with ASTM B 241 (except that the use of alloy 5456 is prohibited), and shall be marked at each end of each length indicating compliance. Aluminum alloy pipe shall be coated to protect against external corrosion where it is in contact with masonry, plaster, or insulation, or is subject to repeated wettings by such liquids as water, detergents, or sewage. Aluminum alloy pipe shall not be used in exterior locations or underground.

403.5 Metallic tubing. Seamless copper, aluminum alloy, or steel tubing shall be permitted to be used with gases not corrosive to such material.

403.5.1 Steel tubing. Steel tubing shall comply with ASTM A 254 or ASTM A 539.

403.5.2 Copper tubing. Copper tubing shall comply with Standard Type K or L of ASTM B 88 or ASTM B 280.

Copper and brass tubing shall not be used if the gas contains more than an average of 0.3 grains of hydrogen sulfide per 100 standard cubic feet of gas (0.7 milligrams per 100 liters).

403.5.3 Aluminum tubing. Aluminum alloy tubing shall comply with ASTM B 210 or ASTM B 241. Aluminum-alloy tubing shall be coated to protect against external corrosion where it is in contact with masonry, plaster, or insulation, or is subject to repeated wettings by such liquids as water, detergent, or sewage.

Aluminum-alloy tubing shall not be used in exterior locations or underground.

403.5.4 Corrugated stainless steel tubing. Corrugated stainless steel tubing shall be tested and listed in compliance with the construction, installation, and performance requirements of ANSI/AGA LC 1.

403.6 Plastic pipe, tubing, and fittings. Plastic pipe, tubing, and fittings shall be used outside underground only and shall conform with ASTM D 2513. Pipe shall be marked "gas" and "ASTM D 2513."

403.6.1 Anodeless risers. Plastic pipe, tubing, anodeless risers shall comply with the following:
1. Factory-assembled anodeless risers shall be recommended by the manufacturer for the gas used and shall be leak tested by the manufacturer in accordance with written procedures.
2. Service head adapters and field-assembled anodeless risers incorporating service head adapters shall be recommended by the manufacturer for the gas used by the manufacturer and shall be designed certified to meet the requirements of Category I of ASTM D 2513, and U.S. Department of Transportation, Code of Federal Regulations, Title 49, Part 192.281(e). The manufacturer shall provide the user qualified installation instructions as prescribed by the U.S. Department of Transportation, Code of Federal Regulations, Title 49, Part 192.283(b).

403.6.2 LP-gas systems. The use of plastic pipe, tubing and fittings in undiluted liquefied petroleum gas piping systems shall be in accordance with NFPA 58.

403.7 Workmanship and defects. Pipe or tubing and fittings shall be clear and free from cutting burrs and defects in structure or threading, and shall be thoroughly brushed, and chip and scale blown.

Defects in pipe, tubing or fittings shall not be repaired. Defective pipe, tubing, and fittings shall be replaced (see Section 406.1.2).

403.8 Protective coating. Where in contact with material or atmosphere exerting a corrosive action, metallic piping and fittings coated with a corrosion-resistant material shall be used. External or internal coatings or linings used on piping or components shall not be considered as adding strength.

403.9 Metallic pipe threads. Metallic pipe and fitting threads shall be taper pipe threads and shall comply with ASME B1.20.1.

403.9.1 Damaged threads. Pipe with threads that are stripped, chipped, corroded, or otherwise damaged shall not be used. If a weld opens during the operation of cutting or threading, that portion of the pipe shall not be used.

403.9.2 Number of threads. Field threading of metallic pipe shall be in accordance with Table 403.9.2.

TABLE 403.9.2
SPECIFICATIONS FOR THREADING METALLIC PIPE

IRON PIPE SIZE (INCHES)	APPROXIMATE LENGTH OF THREADED PORTION	APPROXIMATE NUMBER OF THREADS TO BE CUT
1/2	3/4	10
3/4	3/4	10
1	7/8	10
1 1/4	1	11
1 1/2	1	11
2	1	11
2 1/2	1 1/2	12
3	1 1/2	12
4	1 5/8	13

For SI: 1 inch = 25.4 mm.

403.9.3 Thread compounds. Thread (joint) compounds (pipe dope) shall be resistant to the action of liquefied petroleum gas or to any other chemical constituents of the gases to be conducted through the piping.

403.10 Metallic piping joints and fittings. The type of piping joint used shall be suitable for the pressure-temperature conditions and shall be selected giving consideration to joint tightness and mechanical strength under the service conditions. The joint shall be able to sustain the maximum end force due to the internal pressure and any additional forces due to temperature expansion or contraction, vibration, fatigue, or to the weight of the pipe and its contents.

403.10.1 Pipe joints. Pipe joints shall be threaded, flanged, or welded, and nonferrous pipe shall be permitted to also be brazed with materials having a melting point in excess of 1,000°F (538°C). Brazing alloys shall not contain more than 0.05-percent phosphorus.

403.10.2 Tubing joints. Tubing joints shall be either made with approved gas tubing fittings or brazed with a material having a melting point in excess of 1,000°F (538°C). Brazing alloys shall not contain more than 0.05-percent phosphorus.

403.10.3 Flared joints. Flared joints shall be used only in systems constructed from nonferrous pipe and tubing where experience or tests have demonstrated that the joint is suitable for the conditions and where provisions are made in the design to prevent separation of the joints.

403.10.4 Metallic fittings Metallic fittings, including valves, strainers and filters, shall comply with the following:

1. Threaded fittings in sizes larger than 4 inches (102 mm) shall not be used except where approved.
2. Fittings used with steel or wrought-iron pipe shall be steel, brass, bronze, malleable iron, or cast iron.
3. Fittings used with copper or brass pipe shall be copper, brass, or bronze.
4. Fittings used with aluminum alloy pipe shall be of aluminum alloy.
5. Cast-iron fittings:
   5.1 Flanges shall be permitted to be used.
   5.2 Bushings shall not be used.
   5.3 Fittings shall not be used in systems containing flammable gas-air mixtures.
   5.4 Fittings in sizes 4 inches (102 mm) and larger shall not be used indoors except where approved.
   5.5 Fittings in sizes 6 inches (152 mm) and larger shall not be used except where approved.
6. Brass, bronze, or copper fittings. Fittings, if exposed to soil, shall have a minimum 80-percent copper content.
7. Aluminum alloy fittings. Threads shall not form the joint seal.
8. Zinc-aluminum alloy fittings. Fittings shall not be used in systems containing flammable gas-air mixtures.
9. Special fittings. Fittings such as couplings, proprietary-type joints, saddle tees, gland-type compression fittings, and flared, flareless, or compression-type tubing fittings shall be permitted to be used provided they are used within the fitting manufacturers' pressure-temperature recommendations; used

within the service conditions anticipated with respect to vibration, fatigue, thermal expansion, or contraction; installed or braced to prevent separation of the joint by gas pressure or external physical damage; and approved.

403.11 Plastic piping, joints, and fittings. Plastic pipe, tubing, and fittings shall be joined in accordance with the manufacturers' instructions. Such joint shall comply with the following:

1. The joint shall be designed and installed so that the longitudinal pull-out resistance of the joint will be at least equal to the tensile strength of the plastic piping material.

2. Heat-fusion joints shall be made in accordance with qualified procedures that have been established and proven by test to produce gas-tight joints at least as strong as the pipe or tubing being joined. Joints shall be made with the joining method recommended by the pipe manufacturer. Heat fusion fittings shall be marked "ASTM D 2513."

3. Where compression-type mechanical joints are used, the gasket material in the fitting shall be compatible with the plastic piping and with the gas distributed by the system. An internal tubular rigid stiffener shall be used in conjunction with the fitting. The stiffener shall be flush with the end of the pipe or tubing and shall extend at least to the outside end of the pipe or tubing and at least to the outside end of the compression fitting when installed. The stiffener shall be free of rough or sharp edges and shall not be a force fit in the plastic. Split tubular stiffeners shall not be used.

4. Plastic piping joints and fittings for use in liquefied petroleum gas piping systems shall be in accordance with NFPA 58.

403.12 Flanges. All flanges shall comply with ASME B16.1, ASME B16.20, AWWA C111/A21.11 or MSS SP-6. The pressure-temperature ratings shall equal or exceed that required by the application.

403.12.1 Flange facings. Standard facings shall be permitted for use under this code. Where 150-pound (1034 kPa) pressure rated steel flanges are bolted to Class 125 cast-iron flanges, the raised face on the steel flange shall be removed.

403.12.2 Lapped flanges. Lapped flanges shall be permitted to be used only above ground or in exposed locations accessible for inspection.

403.13 Flange gaskets. Material for gaskets shall be capable of withstanding the design temperature and pressure of the piping system, and the chemical constituents of the gas being conducted, without change to its chemical and physical properties. The effects of fire exposure to the joint shall be considered in choosing material. Acceptable materials include metal or metal-jacketed asbestos (plain or corrugated), asbestos, and aluminum "O" rings and spiral wound metal gaskets. When a flanged joint is opened, the gasket shall be replaced. Full-face gaskets shall be used with all bronze and cast-iron flanges.

SECTION 404
PIPING SYSTEM INSTALLATION

404.1 Prohibited locations. Piping shall not be installed in or through a circulating air duct, clothes chute, chimney or gas vent, ventilating duct, dumbwaiter, or elevator shaft.

404.2 Piping in solid partitions and walls. Concealed piping shall not be located in solid partitions and solid walls, unless installed in a chase or casing.

404.3 Piping in concealed locations. Portions of a piping system installed in concealed locations shall not have unions, tubing fittings, right and left couplings, bushings, compression couplings and swing joints made by combinations of fittings.

Exceptions:
1. Tubing joined by brazing.
2. Fittings listed for use in concealed locations.

404.4 Piping through foundation wall. Underground piping, where installed below grade through the outer foundation or basement wall of a building, shall be encased in a protective pipe sleeve. The annular space between the gas piping and the sleeve shall be sealed.

404.5 Protection against physical damage. In concealed locations, where piping other than black or galvanized steel is installed through holes or notches in wood studs, joists, rafters or similar members less than 1 inch (25 mm) from the nearest edge of the member, the pipe shall be protected by shield plates. Shield plates shall be a minimum of $^1/16$-inch-thick (1.6 mm) steel, shall cover the area of the pipe where the member is notched or bored, and shall extend a minimum of 4 inches (102 mm) above sole plates, below top plates and to each side of a stud, joist or rafter.

404.6 Piping in solid floors. Piping in solid floors shall be laid in channels in the floor and covered in a manner that will allow access to the piping with a minimum amount of damage to the building. Where such piping is subject to exposure to excessive moisture or corrosive substances, the piping shall be protected in an approved manner. As an alternative to installation in channels, the piping shall be installed in a casing of Schedule 40 steel, wrought iron, PVC or ABS pipe with tightly sealed ends and joints. Both ends of such casing shall extend not less than 2 inches (51 mm) beyond the point where the pipe emerges from the floor.

404.7 Above-ground piping outdoors. Piping installed above ground outdoors shall be securely supported and located where it will be protected from physical damage. Where passing through an outside wall, the piping shall also be protected against corrosion by coating or wrapping with an inert material. Where piping is encased in a protective pipe sleeve, the annular space between the piping and the sleeve shall be sealed.

404.8 Protection against corrosion. Metallic pipe or tubing exposed to corrosive action, such as soil condition or moisture, shall be protected in an approved manner. Zinc coatings (galvanizing) shall not be deemed adequate protection for gas piping underground. Ferrous metal exposed in exterior locations shall be protected from corrosion. Where dissimilar metals are joined underground, an insulating coupling or fitting shall be used. Piping shall not be laid in contact with cinders.

404.8.1 Prohibited use. Uncoated threaded or socket welded joints shall not be used in piping in contact with soil or where internal or external crevice corrosion is known to occur.

404.8.2 Protective coatings and wrapping. Pipe protective coatings and wrappings shall be approved for the application and shall be factory applied.

Exception: Where installed in accordance with the manufacturer's installation instructions, field application of coatings and wrappings shall be permitted for pipe nipples, fittings and locations where the factory coating or wrapping has been damaged or necessarily removed at joints.

404.9 Minimum burial depth. Underground piping systems shall be installed a minimum depth of 12 inches (305 mm) below grade, except as provided for in Section 404.9.1.

404.9.1 Individual outside appliances. Individual lines to outside lights, grills or other appliances shall be installed a minimum of 8 inches (203 mm) below finished grade, provided that such installation is approved and is installed in locations not susceptible to physical damage.

404.10 Trenches. The trench shall be graded so that the pipe has a firm, substantially continuous bearing on the bottom of the trench.

404.11 Piping underground beneath buildings. Piping installed underground beneath buildings is prohibited except where the piping is encased in a conduit of wrought iron, plastic pipe, or steel pipe designed to withstand the superimposed loads. Such conduit shall extend into an occupiable portion of the building and, at the point where the conduit terminates in the building, the space between the conduit and

the gas piping shall be sealed to prevent the possible entrance of any gas leakage. Where the end sealing is capable of withstanding the full pressure of the gas pipe, the conduit shall be designed for the same pressure as the pipe. Such conduit shall extend not less than 4 inches (102 mm) outside the building, shall be vented above grade to the outdoors, and shall be installed so as to prevent the entrance of water and insects. The conduit shall be protected from corrosion in accordance with Section 404.8.

404.12 Outlet closures. Gas outlets that do not connect to appliances shall be capped gas tight.

Exception: Listed and labeled flush-mounted-type quick-disconnect devices and listed and labeled gas convenience outlets shall be installed in accordance with the manufacturer's installation instructions.

404.13 Location of outlets. The unthreaded portion of piping outlets shall extend not less than 1 inch (25 mm) through finished ceilings and walls and where extending through floors or outdoor patios and slabs, shall not be less than 2 inches (51 mm) above them. The outlet fitting or piping shall be securely supported. Outlets shall not be placed behind doors. Outlets shall be located in the room or space where the appliance is installed.

Exception: Listed and labeled flush-mounted-type quick-disconnect devices and listed and labeled gas convenience outlets shall be installed in accordance with the manufacturer's installation instructions.

404.14 Plastic pipe. The installation of plastic pipe shall comply with Sections 404.14.1 through 404.14.3.

404.14.1 Limitations. Plastic pipe shall be installed outside underground only. Plastic pipe shall not be used within or under any building or slab or be operated at pressures greater than 100 psig (689 kPa) for natural gas or 30 psig (207 kPa) for LP-gas.

Exceptions:
1. Plastic pipe shall be permitted to terminate aboveground outside of buildings where installed in pre-manufactured anodeless risers or service head adapter risers that are installed in accordance with that manufacturer's installation instructions.
2. Plastic pipe shall be permitted to terminate with a wall head adapter within buildings where the plastic pipe is inserted in a piping material for fuel gas use in buildings.

404.14.2 Connections. Connections made outside and underground between metallic and plastic piping shall be made only with transition fittings categorized as Category I in accordance with ASTM D 2513.

404.14.3 Tracer. A yellow insulated copper tracer wire or other approved conductor shall be installed adjacent to underground nonmetallic piping. Access shall be provided to the tracer wire or the tracer wire shall terminate above ground at each end of the nonmetallic piping. The tracer wire size shall not be less than 18 AWG and the insulation type shall be suitable for direct burial.

404.15 Prohibited devices. A device shall not be placed inside the piping or fittings that will reduce the cross-sectional area or otherwise obstruct the free flow of gas.

Exception: Approved gas filters.

404.16 Testing of piping. Before any system of piping is put in service or concealed, it shall be tested to ensure that it is gas tight. Testing, inspection and purging of piping systems shall comply with Section 406.

SECTION 405
PIPING BENDS AND CHANGES IN DIRECTION

405.1 General. Changes in direction of pipe shall be permitted to be made by the use of fittings, factory bends, or field bends.

405.2 Metallic pipe. Metallic pipe bends shall comply with the following:

1. Bends shall be made only with bending equipment and procedures intended for that purpose.
2. All bends shall be smooth and free from buckling, cracks, or other evidence of mechanical damage.
3. The longitudinal weld of the pipe shall be near the neutral axis of the bend.
4. Pipe shall not be bent through an arc of more than 90 degrees (1.6 rad).
5. The inside radius of a bend shall be not less than 6 times the outside diameter of the pipe.

405.3 Plastic pipe. Plastic pipe bends shall comply with the following:

1. The pipe shall not be damaged and the internal diameter of the pipe shall not be effectively reduced.
2. Joints shall not be located in pipe bends.
3. The radius of the inner curve of such bends shall not be less than 25 times the inside diameter of the pipe.
4. Where the piping manufacturer specifies the use of special bending equipment or procedures, such equipment or procedures shall be used.

405.4 Mitered bends. Mitered bends are permitted subject to the following limitations:

1. Miters shall not be used in systems having a design pressure greater than 50 psig (340 kPa gauge). Deflections caused by misalignments up to 3 degrees

(0.05 rad) shall not be considered as miters.
2. The total deflection angle at each miter shall not exceed 90 degrees (1.6 rad).

405.5 Elbows. Factory-made welding elbows or transverse segments cut therefrom shall have an arc length measured along the crotch at least 1 inch (25 mm) in pipe sizes 2 inches (51 mm) and larger.

SECTION 406
INSPECTION, TESTING AND PURGING

406.1 General. Prior to acceptance and initial operation, all piping installations shall be inspected and pressure tested to determine that the materials, design, fabrication, and installation practices comply with the requirements of this code.

406.1.1 Inspections. Inspection shall consist of visual examination, during or after manufacture, fabrication, assembly, or pressure tests as appropriate. Supplementary types of nondestructive inspection techniques, such as magnetic-particle, radiographic, ultrasonic, etc., shall not be required unless specifically listed herein or in the engineering design.

406.1.2 Repairs and additions. In the event repairs or additions are made following the pressure test, the affected piping shall be tested.

Exception: Minor repairs or additions, provided the work is inspected and connections are tested with a noncorrosive leak-detecting fluid.

406.1.3 Section testing. A piping system shall be permitted to be tested as a complete unit or in sections. Under no circumstances shall a valve in a line be used as a bulkhead between gas in one section of the piping system and test medium in an adjacent section, unless two valves are installed in series with a valved "telltale" located between these valves. A valve shall not be subjected to the test pressure unless it can be determined that the valve, including the valve-closing mechanism, is designed to safely withstand the test pressure.

406.1.4 Regulators and valve assemblies. Regulator and valve assemblies fabricated independently of the piping system in which they are to be installed shall be permitted to be tested with inert gas or air at the time of fabrication.

406.2 Test medium. The test medium shall be air or an inert gas. Oxygen shall not be used.

406.3 Test preparation. Pipe joints, including welds, shall be left exposed for examination during the test. If the pipe end joints have been previously tested in accordance with this code, they shall be permitted to be covered or concealed.

406.3.1 Expansion joints. Expansion joints shall be provided with temporary restraints, if required, for the additional thrust load under test.

406.3.2 Equipment isolation. Equipment that is not to be included in the test shall be either disconnected from the piping or isolated by blanks, blind flanges, or caps. Flanged joints at which blinds are inserted to blank off other equipment during the test shall not be required to be tested.

406.3.3 Equipment disconnection. Where the piping system is connected to equipment or components designed for operating pressures of less than the test pressure, such equipment or equipment components shall be isolated from the piping system by disconnecting them and capping the outlet(s).

406.3.4 Valve isolation. Where the piping system is connected to equipment or components designed for operating pressures equal to or greater than the test pressure, such equipment shall be isolated from the piping system by closing the individual equipment shutoff valve(s).

406.3.5 Testing precautions. All testing of piping systems shall be done with due regard for the safety of employees and the public during the test. Bulkheads, anchorage, and bracing suitably designed to resist test pressures shall be installed if necessary. Prior to testing, the interior of the pipe shall be cleared of all foreign material.

406.4 Test pressure measurement. Test pressure shall be measured with a manometer or with a pressure measuring device designed and calibrated to read, record, or indicate a pressure loss due to leakage during the pressure test period. The source of pressure shall be isolated before the pressure tests are made.

406.4.1 Test pressure. The test pressure to be used shall be no less than $1^{1}/_{2}$ times the proposed maximum working pressure, but not less than 3 psig (20 kPa gauge), irrespective of design pressure. Where the test pressure exceeds 125 psig (862 kPa gauge), the test pressure shall not exceed a value that produces a hoop stress in the piping greater than 50 percent of the specified minimum yield strength of the pipe.

406.4.2 Test duration. Test duration shall be not less than $^{1}/_{2}$ hour for each 500 cubic feet (14 m^3) of pipe volume or fraction thereof. When testing a system having a volume less than 10 cubic feet (0.28 m^3) or a system in a single-family dwelling, the test duration shall be permitted to be reduced to 10 minutes. For piping systems having a volume of more than 24,000 cubic feet (680 m^3), the duration of the test shall not be required to exceed 24 hours.

406.5 Detection of leaks and defects. The piping system shall withstand the test pressure specified without showing any evidence of leakage or other defects.

Any reduction of test pressures as indicated by pressure gages shall be deemed to indicate the presence of a leak unless such reduction can be readily attributed to some other cause.

406.5.1 Detection methods. The leakage shall be located by means of an approved combustible gas detector, a noncorrosive leak detection fluid, or an equivalent nonflammable solution. Matches, candles, open flames, or other methods that could provide a source of ignition shall not be used.

406.5.2 Corrections. Where leakage or other defects are located, the affected portion of the piping system shall be repaired or replaced and retested.

406.6 System and equipment leakage test. Leakage testing of systems and equipment shall be in accordance with Sections 406.6.1 through 406.6.4.

406.6.1 Test gases. Fuel gas shall be permitted to be used for leak checks in piping systems that have been tested in accordance with Section 406.

406.6.2 Before turning gas on. Before gas is introduced into a system of new gas piping, the entire system shall be inspected to determine that there are no open fittings or ends and that all manual valves at outlets on equipment are closed and all unused valves at outlets are closed and plugged or capped.

406.6.3 Test for leakage. Immediately after the gas is turned on into a new system or into a system that has been initially restored after an interruption of service, the piping system shall be tested for leakage. If leakage is indicated, the gas supply shall be shut off until the necessary repairs have been made.

406.6.4 Placing equipment in operation. Gas utilization equipment shall be permitted to be placed in operation after the piping system has been tested and determined to be free of leakage and purged in accordance with Section 406.7.2.

406.7 Purging. Purging of piping shall comply with Sections 406.7.1 through 406.7.4.

406.7.1 Removal from service. Where gas piping is to be opened for servicing, addition, or modification, the section to be worked on shall be turned off from the gas supply at the nearest convenient point, and the line pressure vented to the outdoors, or to ventilated areas of sufficient size to prevent accumulation of flammable mixtures.

The remaining gas in this section of pipe shall be displaced with an inert gas as required by Table 406.7.1.

TABLE 406.7.1
LENGTH OF PIPING REQUIRING PURGING WITH INERT GAS FOR SERVICING OR MODIFICATION

NOMINAL PIPE SIZE, (inches)	LENGTH OF PIPING REQUIRING PURGING
$2^1/_2$	> 50 feet
3	> 30 feet
4	> 15 feet
6	> 10 feet
8 or larger	Any length

For SI: 1 inch = 25.4 mm, 1 foot = 304.8 mm.

406.7.2 Placing in operation. Where piping full of air is placed in operation, the air in the piping shall be displaced with fuel gas, provided the piping does not exceed the length shown in Table 406.7.2. The air can be safely displaced with fuel gas provided that a moderately rapid and continuous flow of fuel gas is introduced at one end of the line and air is vented out at the other end. The fuel gas flow shall be continued without interruption until the vented gas is free of air. The point of discharge shall not be left unattended during purging. After purging, the vent shall then be closed. Where required by Table 406.7.2, the air in the piping shall first be displaced with an inert gas, and the inert gas shall then be displaced with fuel gas.

TABLE 406.7.2
LENGTH OF PIPING REQUIRING PURGING WITH INERT GAS BEFORE PLACING IN OPERATION

NOMINAL PIPE SIZE, (inches)	LENGTH OF PIPING REQUIRING PURGING
3	> 30 feet
4	> 15 feet
6	> 10 feet
8 or larger	Any length

For SI: 1 inch = 25.4 mm, 1 foot = 304.8 mm.

406.7.3 Discharge of purged gases. The open end of piping systems being purged shall not discharge into confined spaces or areas where there are sources of ignition unless precautions are taken to perform this operation in a safe manner by ventilation of the space, control of purging rate, and elimination of all hazardous conditions.

406.7.4 Placing equipment in operation. After the piping has been placed in operation, all equipment shall be purged and then placed in operation, as necessary.

SECTION 407
PIPING SUPPORT

407.1 General. Piping shall be provided with support in accordance with Section 407.2.

407.2 Design and installation. Piping shall be supported with pipe hooks, metal pipe straps, bands, brackets, or hangers suitable for the size of piping, of adequate strength and quality, and located at intervals so as to prevent or damp out excessive vibration. Piping shall be anchored to prevent undue strains on connected equipment and shall not be supported by other piping. Pipe hangers and supports shall conform to the requirements of MSS SP-58 and shall be spaced in accordance with Section 415. Supports, hangers, and anchors shall be installed so as not to interfere with the free expansion and contraction of the piping between anchors. All parts of the supporting equipment shall be designed and installed so they will not be disengaged by movement of the supported piping.

SECTION 408
DRIPS AND SLOPED PIPING

408.1 Slopes. Piping for other than dry gas conditions shall be sloped not less than $^1/_4$ inch in 15 feet (6.3 mm in 4572 mm) to prevent traps.

408.2 Drips. Where wet gas exists, a drip shall be provided at any point in the line of pipe where condensate could collect. A drip shall also be provided at the outlet of the meter and shall be installed so as to constitute a trap wherein an accumulation of condensate will shut off the flow of gas before the condensate will run back into the meter.

408.3 Location of drips. Drips shall be provided with ready access to permit cleaning or emptying. A drip shall not be located where the condensate is subject to freezing.

408.4 Sediment trap. Where a sediment trap is not incorporated as a part of the gas utilization equipment, a sediment trap shall be installed as close to the inlet of the equipment as practical. The sediment trap shall be either a tee fitting with a capped nipple in the bottom opening of the run of the tee or other device approved as an effective sediment trap. Illuminating appliances, ranges, clothes dryers, and outdoor grills need not be so equipped.

SECTION 409
SHUTOFF VALVES

409.1 General. Piping systems shall be provided with shutoff valves in accordance with this section.

409.1.1 Valve approval. Shutoff valves shall be of an approved type. Shutoff valves shall be constructed of materials compatible with the piping. Shutoff valves installed in a portion of a piping system operating above 0.5 psig shall comply with ASME B16.33. Shutoff valves installed in a portion of a piping system operating at 0.5 psig or less shall comply with ANSI Z21.15 or ASME B16.33.

409.1.2 Prohibited locations. Shutoff valves shall be prohibited in concealed locations and spaces used as plenums.

409.1.3 Access to shutoff valves. Shutoff valves shall be located in places so as to provide access for operation and shall be installed so as to be protected from damage.

409.2 Meter valve. Every meter shall be equipped with a shutoff valve located on the supply side of the meter.

409.3 Shutoff valves for multiple-house line systems. Where a single meter is used to supply gas to more than one building or tenant, a separate shutoff valve shall be provided for each building or tenant.

409.3.1 Multiple tenant buildings. In multiple tenant buildings, where a common piping system is installed to supply other than one- and two-family dwellings, shutoff valves shall be provided for each tenant. Each tenant shall have access to the shutoff valve serving that tenant's space.

409.3.2 Individual buildings. In a common system serving more than one building, shutoff valves shall be installed outdoors at each building.

409.3.3 Identification of shutoff valves. Each house line shutoff valve shall be plainly marked with an identification tag attached by the installer so that the piping systems supplied by such valves are readily identified.

409.4 MP Regulator valves. A listed shutoff valve shall be installed immediately ahead of each MP regulator.

409.5 Equipment shutoff valve. Each appliance shall be provided with a shutoff valve separate from the appliance. The shutoff valve shall be located in the same room as the appliance, not further than 6 feet (1829 mm) from the appliance, and shall be installed upstream from the union, connector or quick disconnect device it serves. Such shutoff valves shall be provided with ready access.

Exception: Shutoff valves for vented decorative appliances and decorative appliances for installation in vented fireplaces shall not be prohibited from being installed in an area remote from the appliance where such valves are provided with ready access. Such valves shall be permanently identified and shall serve no other equipment.

409.5.1 Shutoff valve in fireplace. Equipment shutoff valves located in the firebox of a fireplace shall be installed in accordance with the appliance manufacturer's instructions.

SECTION 410
FLOW CONTROLS

410.1 Pressure regulators. A line pressure regulator shall be installed where the appliance is designed to operate at a lower pressure than the supply pressure. Access shall be provided to pressure regulators. Pressure regulators shall be protected from physical damage. Regulators installed on the exterior of the building shall be approved for outdoor installation.

410.2 MP regulators. MP pressure regulators shall comply with the following:

1. The MP regulator shall be approved and shall be suitable for the inlet and outlet gas pressures for the application.
2. The MP regulator shall maintain a reduced outlet pressure under lockup (no-flow) conditions.
3. The capacity of the MP regulator, determined by published ratings of its manufacturer, shall be adequate to supply the appliances served.
4. The MP pressure regulator shall be provided with access. Where located indoors, the regulator shall be vented to the outdoors or shall be equipped with a leak-limiting device, in either case complying with Section 410.3.
5. A tee fitting with one opening capped or plugged shall be installed between the MP regulator and its upstream shutoff valve. Such tee fitting shall be positioned to allow connection of a pressure-measuring instrument and to serve as a sediment trap.
6. A tee fitting with one opening capped or plugged shall be installed not less than 10 pipe diameters downstream of the MP regulator outlet. Such tee fitting shall be positioned to allow connection of a pressure measuring instrument.

410.3 Venting of regulators. Pressure regulators that require a vent shall have an independent vent to the outside of the building. The vent shall be designed to prevent the entry of water or foreign objects.

Exception: A vent to the outside of the building is not required for regulators equipped with and labeled for utilization with approved vent-limiting devices installed in accordance with the manufacturer's instructions.

SECTION 411
APPLIANCE CONNECTIONS

411.1 Connecting appliances. Appliances shall be connected to the piping system by one of the following:
1. Rigid metallic pipe and fittings.
2. Semirigid metallic tubing and metallic fittings. Lengths shall not exceed 6 feet (1829 mm) and shall be located entirely in the same room as the appliance. Semirigid metallic tubing shall not enter a motor-operated appliance through an unprotected knockout opening.
3. Listed and labeled appliance connectors installed in accordance with the manufacturer's installation instructions and located entirely in the same room as the appliance.
4. Listed and labeled quick-disconnect devices used in conjunction with listed and labeled appliance connectors.
5. Listed and labeled convenience outlets used in conjunction with listed and labeled appliance connectors.
6. Listed and labeled appliance connectors complying with ANSI Z21.69 and listed for use with food service equipment having casters, or that is otherwise subject to movement for cleaning, and other large movable equipment.

411.1.1 Protection from damage. Connectors and tubing shall be installed so as to be protected against physical damage.

411.1.2 Appliance fuel connectors. Connectors shall have an overall length not to exceed 3 feet (914 mm), except for range and domestic clothes dryer connectors, which shall not exceed 6 feet (1829 mm) in length. Connectors shall not be concealed within, or extended through, walls, floors, partitions, ceilings or appliance housings. A shutoff valve not less than the nominal size of the connector shall be installed ahead of the connector in accordance with Section 409.5. Connectors shall be sized to provide the total demand of the connected appliance.

Exception: Fireplace inserts factory equipped with grommets, sleeves, or other means of protection in accordance with the listing of the appliance.

411.1.3 Movable appliances. Where appliances are equipped with casters or are otherwise subject to periodic movement or relocation for purposes such as routine cleaning and maintenance, such appliances shall be connected to the supply system piping by means of an approved flexible connector designed and labeled for the application. Such flexible connectors shall be installed and protected against physical damage in accordance with the manufacturer's installation instructions.

SECTION 412
LIQUEFIED PETROLEUM GAS MOTOR VEHICLE FUEL-DISPENSING STATIONS

412.1 General. Service stations for LP-gas fuel shall be in accordance with this section and the *Fire Code of New York State*. The operation of LP-gas service stations shall be regulated by the *Fire Code of New York State*.

412.2 Storage and dispensing. Storage vessels and equipment used for the storage or dispensing of LP-gas shall be approved or listed in accordance with Sections 412.3 and 412.4.

412.3 Approved equipment. Containers; pressure-relief devices, including pressure-relief valves; and pressure regulators and piping used for LP-gas shall be approved.

412.4 Listed equipment. Hoses, hose connections, vehicle fuel connections, dispensers, LP-gas pumps and electrical equipment used for LP-gas shall be listed.

412.5 Attendants. Motor vehicle fueling operations shall be conducted by qualified attendants or in accordance with Section 412.8 by persons trained in the proper handling of LP-gas.

412.6 Location. In addition to the fuel dispensing requirements of the fire code, the point of transfer for dispensing operations shall be 25 feet (7620 mm) or more from buildings having combustible exterior wall surfaces, buildings having noncombustible exterior wall surfaces that are not part of a 1-hour fire-resistant assembly or buildings having combustible overhangs, property which could be built on, public streets, or sidewalks and railroads; and at least 10 feet (3048 mm) from driveways and buildings having noncombustible exterior wall surfaces that are part of a fire-resistant assembly having a rating of 1 hour or more.

Exception: The point of transfer for dispensing operations need not be separated from canopies providing weather protection for the dispensing equipment constructed in accordance with the *Building Code of New York State*.

Liquefied petroleum gas containers shall be located in accordance with the *Fire Code of New York State*. Liquefied petroleum gas storage and dispensing equipment shall be located outdoors and in accordance with the *Fire Code of New York State*.

412.7 Installation of dispensing devices and equipment. The installation and operation of LP-gas dispensing systems shall be in accordance with this section and the *Fire Code of New York State*. Liquefied petroleum gas dispensers and dispensing stations shall be installed in accordance with manufacturer's specifications and their listing.

412.7.1 Valves. A manual shutoff valve and an excess flow-control check valve shall be located in the liquid line between the pump and the dispenser inlet where the dispensing device is installed at a remote location and is not part of a complete storage and dispensing unit mounted on a common base.

An excess flow-control check valve or an emergency shutoff valve shall be installed in or on the dispenser at the point at which the dispenser hose is connected to the liquid piping. A differential backpressure valve shall be considered equivalent protection. A listed shutoff valve shall be located at the discharge end of the transfer hose.

412.7.2 Hoses. Hoses and piping for the dispensing of LP-gas shall be provided with hydrostatic relief valves. The hose length shall not exceed 18 feet (5486 mm). An approved method shall be provided to protect the hose against mechanical damage.

412.7.3 Vehicle impact protection. Vehicle impact protection for LP-gas storage containers, pumps and dispensers shall be provided in accordance with the *Fire Code of New York State*.

412.8 Private fueling of motor vehicles. Self-service LP-gas dispensing systems, including key, code and card lock dispensing systems, shall not be open to the public and shall be limited to the filling of permanently mounted fuel containers on LP-gas powered vehicles. In addition to the requirements in the *Fire Code of New York State*, self-service LP-gas dispensing systems shall be provided with an emergency shutoff switch located within 100 feet (30 480 mm) of, but not less than 20 feet (6096 mm) from, dispensers and the owner of the dispensing facility shall ensure the safe operation of the system and the training of users.

SECTION 413
COMPRESSED NATURAL GAS MOTOR VEHICLE FUEL-DISPENSING STATIONS

413.1 General. Service stations for CNG fuel shall be in accordance with this section and the *Fire Code of New York State*. The operation of CNG service stations shall be regulated by the *Fire Code of New York State*.

413.2 General. Storage vessels and equipment used for the storage, compression or dispensing of CNG shall be approved or listed in accordance with Sections 413.2.1 and 413.2.2.

413.2.1 Approved equipment. Containers; compressors; pressure-relief devices, including pressure-relief valves; and pressure regulators and piping used for CNG shall be approved.

413.2.2 Listed equipment. Hoses, hose connections, dispensers, gas-detection systems and electrical equipment used for CNG shall be listed. Vehicle fueling connections shall be listed and labeled.

413.3 Location of dispensing operations and equipment. Compression, storage and dispensing equipment shall be located above ground outside.

Exceptions:
1. Compression, storage or dispensing equipment is allowed in buildings of noncombustible construction, as set forth in the *Building Code of New York State*, which are unenclosed for three-quarters or more of the perimeter.
2. Compression, storage and dispensing equipment is allowed to be located indoors in accordance with the *Fire Code of New York State*.

413.3.1 Location on property. In addition to the fuel-dispensing requirements of the *Fire Code of New York State*, compression, storage and dispensing equipment shall not be installed:

1. Beneath power lines,
2. Less than 10 feet (3048 mm) from the nearest building or property line which could be built on, public street, sidewalk, or source of ignition.

   Exception: Dispensing equipment need not be separated from canopies providing weather protection for the dispensing equipment constructed in accordance with the *Building Code of New York State*.

3. Less than 25 feet (7620 mm) from the nearest rail of any railroad track.
4. Less than 50 feet (15 240 mm) from the nearest rail of any railroad main track or any railroad or transit line where power for train propulsion is provided by an outside electrical source such as third rail or overhead catenary.
5. Less than 50 feet (15 240 mm) from the vertical plane below the nearest overhead wire of a trolley bus line.

413.4 Private fueling of motor vehicles. Self-service CNG-dispensing systems, including key, code and card lock dispensing systems, shall be limited to the filling of permanently mounted fuel containers on CNG-powered vehicles.

In addition to the requirements in the *Fire Code of New York State*, the owner of a self-service CNG-dispensing facility shall ensure the safe operation of the system and the training of users.

413.5 Pressure regulators. Pressure regulators shall be designed, installed or protected so their operation will not be affected by the elements (freezing rain, sleet, snow, ice, mud or debris). This protection is allowed to be integral with the regulator.

413.6 Valves. Piping to equipment shall be provided with a manual shutoff valve. Such valve shall be provided with ready access.

413.7 Emergency shutdown equipment. An emergency shutdown device shall be located within 75 feet (22 860 mm) of, but not less than 25 feet (7620 mm) from, dispensers and shall also be provided in the compressor area. Upon activation, the emergency shutdown shall automatically shut off the power supply to the compressor and close valves between the main gas supply and the compressor and between the storage containers and dispensers.

413.8 Discharge of CNG from motor vehicle fuel storage containers. The discharge of CNG from motor vehicle fuel cylinders for the purposes of maintenance, cylinder certification, calibration of dispensers or other activities shall be in accordance with this section. The discharge of CNG from motor vehicle fuel cylinders shall be accomplished through a closed transfer system or an approved method of atmospheric venting in accordance with Section 413.8.1 or 413.8.2.

413.8.1 Closed transfer system. A documented procedure which explains the logical sequence for discharging the cylinder shall be provided to the code enforcement official. The procedure shall include what actions the operator will take in the event of a low-pressure or high-pressure natural gas release during the discharging activity. A drawing illustrating the arrangement of piping, regulators and equipment settings shall be provided to the code enforcement official. The drawing shall illustrate the piping and regulator arrangement and shall be shown in spatial relation to the location of the compressor, storage vessels and emergency shutdown devices.

413.8.2 Atmospheric venting. Atmospheric venting of motor vehicle fuel cylinders shall be in accordance with Sections 413.8.2.1 through 413.8.2.6.

413.8.2.1 Plans and specifications. A drawing illustrating the location of the vessel support, piping, the method of grounding and bonding, and other requirements specified herein shall be provided to the code enforcement official.

413.8.2.2 Cylinder stability. A method of rigidly supporting the vessel during the venting of CNG shall be provided. The selected method shall provide not less than two points of support and shall prevent the hori-

zontal and lateral movement of the vessel. The system shall be designed to prevent the movement of the vessel based on the highest gas-release velocity through valve orifices at the vessel's rated pressure and volume. The structure or appurtenance shall be constructed of noncombustible materials.

413.8.2.3 Separation. The structure or appurtenance used for stabilizing the cylinder shall be separated from the site equipment, features and exposures and shall be located in accordance with Table 413.8.2.3.

TABLE 413.8.2.3
SEPARATION DISTANCE FOR
ATMOSPHERIC VENTING OF CNG

EQUIPMENT OR FEATURE	MINIMUM SEPARATION (feet)
Buildings	25
Building openings	25
Lot lines	15
Public ways	15
Vehicles	25
CNG compressor and and storage vessels	25
CNG dispensers	25

For SI: 1 foot = 304.8 mm.

413.8.2.4 Grounding and bonding. The structure or appurtenance used for supporting the cylinder shall be grounded in accordance with Chapter 27 of the Building Code of New York State. The cylinder valve shall be bonded prior to the commencement of venting operations.

413.8.2.5 Vent tube. A vent tube that will divert the gas flow to the atmosphere shall be installed on the cylinder prior to the commencement of the venting and purging operation. The vent tube shall be constructed of pipe or tubing materials approved for use with CNG in accordance with the *Fire Code of New York State*.

The vent tube shall be capable of dispersing the gas a minimum of 10 feet (3048 mm) above grade level. The vent tube shall not be provided with a rain cap or other feature which would limit or obstruct the gas flow.

At the connection fitting of the vent tube and the CNG cylinder, a listed bidirectional detonation flame arrester shall be provided.

413.8.2.6 Signage. Approved NO SMOKING signs shall be posted within 10 feet (3048 mm) of the cylinder support structure or appurtenance. Approved CYLINDER SHALL BE BONDED signs shall be posted on the cylinder support structure or appurtenance.

SECTION 414
SUPPLEMENTAL AND STANDBY GAS SUPPLY

414.1 Use of air or oxygen under pressure. Where air or oxygen under pressure is used in connection with the gas supply, effective means such as a backpressure regulator and relief valve shall be provided to prevent air or oxygen from passing back into the gas piping. Where oxygen is used, installation shall be in accordance with NFPA 51.

414.2 Interconnections for standby fuels. Where supplementary gas for standby use is connected downstream from a meter or a service regulator where a meter is not provided, a device to prevent backflow shall be installed. A three-way valve installed to admit the standby supply and at the same time shut off the regular supply shall be permitted to be used for this purpose.

SECTION 415
PIPING SUPPORT INTERVALS

415.1 Interval of support. Piping shall be supported at intervals not exceeding the spacing specified in Table 415.1.

TABLE 415.1
SUPPORT OF PIPING

STEEL PIPE, NOMINAL SIZE OF PIPE (inches)	SPACING OF SUPPORTS (feet)	NOMINAL SIZE OF TUBING (inch O.D.)	SPACING OF SUPPORTS (feet)
$1/2$	6	$1/2$	4
$3/4$ or 1	8	$5/8$ or $3/4$	6
$1^1/4$ or larger (horizontal)	10	$7/8$ or 1	8
$1^1/4$ or larger (vertical)	every floor level		

For SI: 1 inch = 25.4 mm; 1 foot = 304.8 mm.

CHAPTER 5
CHIMNEYS AND VENTS

SECTION 501
GENERAL

501.1 Scope. This chapter shall govern the installation, maintenance, repair and approval of factory-built chimneys, chimney liners, vents and connectors and the utilization of masonry chimneys serving gas-fired appliances. The requirements for the installation, maintenance, repair and approval of factory-built chimneys, chimney liners, vents and connectors serving appliances burning fuels other than fuel gas shall be regulated by the *Mechanical Code of New York State*. The construction, repair, maintenance and approval of masonry chimneys shall be regulated by the *Building Code of New York State*.

501.1.1 Commercial cooking appliances vented by exhaust hoods. Where commercial cooking appliances are vented by means of the Type I or Type II kitchen exhaust hood system that serves such appliances, the appliances shall be interlocked with the exhaust hood system to prevent appliance operation when the exhaust hood system is not operating. Where automatically operated appliances such as water heaters are vented through natural draft kitchen exhaust hoods, dampers shall not be installed in the exhaust system.

Exception: An interlock between the cooking appliance and the exhaust hood system shall not be required for appliances that are of the manually operated type and are factory equipped with standing pilot burner ignition systems.

501.2 General. Every appliance shall discharge the products of combustion to the outdoors, except for appliances exempted by Section 501.8.

501.3 Masonry chimneys. Masonry chimneys shall be constructed in accordance with Section 503.5.3 and the *Building Code of New York State*.

501.4 Minimum size of chimney or vent. Chimneys and vents shall be sized in accordance with Section 504.

501.5 Abandoned inlet openings. Abandoned inlet openings in chimneys and vents shall be closed by an approved method.

501.6 Positive pressure. Where an appliance equipped with a mechanical forced draft system creates a positive pressure in the venting system, the venting system shall be designed for positive pressure applications.

501.7 Connection to fireplace. Connection of appliances to chimney flues serving fireplaces shall be in accordance with Sections 501.7.1 through 501.7.3.

501.7.1 Closure and access. A noncombustible seal shall be provided below the point of connection to prevent entry of room air into the flue. Means shall be provided for access to the flue for inspection and cleaning.

501.7.2 Connection to factory-built fireplace flue. An appliance shall not be connected to a flue serving a factory-built fireplace unless the appliance is specifically listed for such installation. The connection shall be made in accordance with the appliance manufacturer's installation instructions.

501.7.3 Connection to masonry fireplace flue. A connector shall extend from the appliance to the flue serving a masonry fireplace such that the flue gases are exhausted directly into the flue. The connector shall be accessible or removable for inspection and cleaning of both the connector and the flue. Listed direct connection devices shall be installed in accordance with their listing.

501.8 Equipment not required to be vented. The following appliances shall not be required to be vented.

1. Ranges.
2. Built-in domestic cooking units listed and marked for optional venting.
3. Hot plates and laundry stoves.
4. Type 1 clothes dryers (Type 1 clothes dryers shall be exhausted in accordance with the requirements of Section 613.)
5. A single booster-type automatic instantaneous water heater, where designed and used solely for the sanitizing rinse requirements of a dishwashing machine, provided that the heater is installed in a commercial kitchen having a mechanical exhaust system. Where installed in this manner, the draft hood, if required, shall be in place and unaltered and the draft hood outlet shall be not less than 36 inches (914 mm) vertically and 6 inches (152 mm) horizontally from any surface other than the heater.
6. Refrigerators.
7. Counter appliances.
8. Room heaters listed for unvented use.
9. Direct-fired make-up air heaters.
10. Other equipment listed for unvented use and not provided with flue collars.
11. Specialized equipment of limited input such as laboratory burners and gas lights.

Where the appliances and equipment listed in Items 1 through 11 above are installed so that the aggregate input rating exceeds 20 Btu per hour per cubic foot (207 watts per m³) of volume of the room or space in which such appliances and equipment are installed, one or more shall be provided with venting systems or other approved means for conveying the vent gases to the outdoor atmosphere so that the aggregate input rating of the remaining unvented appliances and equipment does not exceed the 20 Btu per hour per cubic foot (207 watts per m³) figure. Where the room or space in which the equipment is installed is directly connected to another room or space by a doorway, archway, or other opening of comparable size that cannot be closed, the volume of such adjacent room or space shall be permitted to be included in the calculations.

501.9 Chimney entrance. Connectors shall connect to a masonry chimney flue at a point not less than 12 inches (305 mm) above the lowest portion of the interior of the chimney flue.

501.10 Connections to exhauster. Appliance connections to a chimney or vent equipped with a power exhauster shall be made on the inlet side of the exhauster. Joints on the positive pressure side of the exhauster shall be sealed to prevent flue-gas leakage as specified by the manufacturer's installation instructions for the exhauster.

501.11 Masonry chimneys. Masonry chimneys utilized to vent appliances shall be located, constructed and sized as specified in the manufacturer's installation instructions for the appliances being vented and Section 503.

501.12 Residential and low-heat appliances flue lining systems. Flue lining systems for use with residential-type and low-heat appliances shall be limited to the following:

1. Clay flue lining complying with the requirements of ASTM C 315 or equivalent. Clay flue lining shall be installed in accordance with the *Building Code of New York State*.
2. Listed chimney lining systems complying with UL 1777.
3. Other approved materials that will resist, without cracking, softening or corrosion, flue gases and condensate at temperatures up to 1,800°F (982°C).

501.13 Category I appliance flue lining systems. Flue lining systems for use with Category I appliances shall be limited to the following:

1. Flue lining systems complying with Section 501.12.
2. Chimney lining systems listed and labeled for use with gas appliances with draft hoods and other Category I gas appliances listed and labeled for use with Type B vents.

501.14 Category II, III and IV appliance venting systems. The design, sizing and installation of vents for Category II, III and IV appliances shall be in accordance with the appliance manufacturer's installation instructions.

501.15 Existing chimneys and vents. Where an appliance is permanently disconnected from an existing chimney or vent, or where an appliance is connected to an existing chimney or vent during the process of a new installation, the chimney or vent shall comply with Sections 501.15.1 through 501.15.4.

501.15.1 Size. The chimney or vent shall be resized as necessary to control flue gas condensation in the interior of the chimney or vent and to provide the appliance or appliances served with the required draft. For Category I appliances, the resizing shall be in accordance with Section 502.

501.15.2 Flue passageways. The flue gas passageway shall be free of obstructions and combustible deposits and shall be cleaned if previously used for venting a solid or liquid fuel-burning appliance or fireplace. The flue liner, chimney inner wall or vent inner wall shall be continuous and shall be free of cracks, gaps, perforations or other damage or deterioration which would allow the escape of combustion products, including gases, moisture and creosote.

501.15.3 Cleanout. Masonry chimney flues shall be provided with a cleanout opening having a minimum height of 6 inches (152 mm). The upper edge of the opening shall be located not less than 6 inches (152 mm) below the lowest chimney inlet opening. The cleanout shall be provided with a tight-fitting, noncombustible cover.

501.15.4 Clearances. Chimneys and vents shall have airspace clearance to combustibles in accordance with the *Building Code of New York State* and the chimney or vent manufacturer's installation instructions. Noncombustible firestopping or fireblocking shall be provided in accordance with the *Building Code of New York State*.

> **Exception:** Masonry chimneys equipped with a chimney lining system tested and listed for installation in chimneys in contact with combustibles in accordance with UL 1777, and installed in accordance with the manufacturer's instructions, shall not be required to have clearance between combustible materials and exterior surfaces of the masonry chimney.

SECTION 502
VENTS

502.1 General. All vents, except as provided in Section 503.7, shall be listed and labeled. Type B and BW vents shall be tested in accordance with UL 441. Type L vents shall be tested in accordance with UL 641. Vents for Category II and III appliances shall be tested in accordance with UL 1738. Plastic vents for Category IV appliances shall not be required to be listed and labeled where such vents are as specified by the appliance manufacturer and are installed in accordance with the appliance manufacturer's installation instructions.

502.2 Connectors required. Connectors shall be used to connect appliances to the vertical chimney or vent, except where the chimney or vent is attached directly to the appliance. Vent connector size, material, construction and installation shall be in accordance with Section 503.

502.3 Vent application. The application of vents shall be in accordance with Table 503.4.

502.4 Insulation shield. Where vents pass through insulated assemblies, an insulation shield constructed of not less than 26 gage sheet (0.016 inch) (0.4 mm) metal shall be installed to provide clearance between the vent and the insulation material. The clearance shall not be less than the clearance to combustibles specified by the vent manufacturer's installation instructions. Where vents pass through attic space, the shield shall terminate not less than 2 inches (51 mm) above the insulation materials and shall be secured in place to prevent displacement. Insulation shields provided as part of a listed vent system shall be installed in accordance with the manufacturer's installation instructions.

502.5 Installation. Vent systems shall be sized, installed and terminated in accordance with the vent and appliance manufacturer's installation instructions and Section 503.

502.6 Support of vents. All portions of vents shall be adequately supported for the design and weight of the materials employed.

SECTION 503
VENTING OF EQUIPMENT

503.1 General. This section recognizes that the choice of venting materials and the methods of installation of venting systems are dependent on the operating characteristics of the equipment being vented. The operating characteristics of vented equipment can be categorized with respect to (1) positive or negative pressure within the venting system; and (2) whether or not the equipment generates flue or vent gases that may condense in the venting system. See Section 202 for the definition of these vented appliance categories.

503.2 Venting systems required. Except as permitted in Sections 503.2.1 through 503.2.4 and 501.8, all equipment shall be connected to venting systems.

503.2.1 Ventilating hoods. Ventilating hoods and exhaust systems shall be permitted to be used to vent equipment installed in commercial applications (see Section 503.3.4) and to vent industrial equipment, such as where the process itself requires fume disposal.

503.2.2 Well-ventilated spaces. Where located in a large and well-ventilated space, industrial equipment shall be permitted to be operated by discharging the flue gases directly into the space.

503.2.3 Direct-vent equipment. Listed direct-vent equipment shall be considered properly vented where installed in accordance with the terms of its listing, the manufacturer's instructions, and Section 503.8(3).

503.2.4 Equipment with integral vents. Equipment incorporating integral venting means shall be considered properly vented when installed in accordance with its listing, the manufacturer's instructions, and Sections 503.8(1) and 503.8(2).

503.3 Design and construction. A venting system shall be designed and constructed so as to develop a positive flow adequate to convey flue or vent gases to the outdoor atmosphere.

503.3.1 Equipment draft requirements. A venting system shall satisfy the draft requirements of the equipment in accordance with the manufacturer's instructions.

503.3.2 Design and construction. Gas utilization equipment required to be vented shall be connected to a venting system designed and constructed in accordance with the provisions of Sections 503.4 through 503.15.

503.3.3 Mechanical draft systems. Mechanical draft systems shall comply with the following:
1. Equipment, except incinerators, requiring venting shall be permitted to be vented by means of mechanical draft systems of either forced or induced draft design.
2. Forced draft systems and all portions of induced draft systems under positive pressure during operation shall be designed and installed so as to prevent leakage of flue or vent gases into a building.
3. Vent connectors serving equipment vented by natural draft shall not be connected into any portion of mechanical draft systems operating under positive pressure.
4. When a mechanical draft system is employed, provision shall be made to prevent the flow of gas to the main burners when the draft system is not performing so as to satisfy the operating requirements of the equipment for safe performance.
5. The exit terminals of mechanical draft systems shall be not less than 7 feet (2134 mm) above grade where located adjacent to public walkways and shall be located as specified in Section 503.8, Items 1 and 2.
6. Mechanical draft systems shall be installed in accordance with the terms of their listing and the manufacturer's instructions.

503.3.4 Ventilating hoods and exhaust systems. Ventilating hoods and exhaust systems shall be permitted to be used to vent gas utilization equipment installed in commercial applications. Where automatically operated equipment is vented through a ventilating hood or exhaust system equipped with a damper or with a power means of exhaust, provisions shall be made to allow the flow of gas to the main burners only when the damper is open to a position to properly vent the equipment and when the power means of exhaust is in operation.

503.3.5 Circulating air ducts and plenums. No portion of a venting system shall extend into or pass through any circulating air duct or plenum.

503.4 Type of venting system to be used. The type of venting system to be used shall be in accordance with Table 503.4.

TABLE 503.4
TYPE OF VENTING SYSTEM TO BE USED

GAS UTILIZATION EQUIPMENT	TYPE OF VENTING SYSTEM
Listed Category I equipment Listed equipment equipped with draft hood Equipment listed for use with Type B gas vent	Type B gas vent (Section 503.6) Chimney (Section 503.5) Single-wall metal pipe (Section 503.7) Listed Chimney lining system for gas venting. (Section 503.5.3) Special gas vent listed for this equipment (Section 503.4.2)
Listed vented wall furnaces	Type B-W gas vent (Sections 503.6, 607)
Category II equipment	As specified or furnished by manufacturers of listed equipment (Sections 503.4.1, 503.4.2)
Category III equipment	As specified or furnished by manufacturers of listed equipment (Sections 503.4.1, 503.4.2)
Category IV equipment	As specified or furnished by manufacturers of listed equipment (Sections 503.4.1, 503.4.2)
Incinerators, indoors	Chimney (Section 503.5)
Incinerators, outdoors	Single-wall metal pipe (Sections 503.7, 503.7.6)
Equipment which may be converted to use of solid fuel	Chimney (Section 503.5)
Unlisted combination gas and oil-burning equipment	Chimney (Section 503.5)
Listed combination gas and oil-burning equipment	Type L vent (Section 503.6) or chimney (Section 503.5)
Combination gas and solid fuel-burning equipment	Chimney (Section 503.5)
Equipment listed for use with chimneys only	Chimney (Section 503.5)
Unlisted equipment	Chimney (Section 503.5)
Decorative appliance in vented fireplace	Chimney
Gas-fired toilets	Single-wall metal pipe (Section 625)
Direct vent equipment	See Section 503.2.3
Equipment with integral vent	See Section 503.2.4
Equipment in commercial and industrial installations	Chimney, ventilating hood, and exhaust system (Section 503.3.4)

503.4.1 Plastic piping. Approved plastic piping shall be permitted to be used for venting equipment listed for use with such venting materials.

503.4.2 Special gas vent. Special gas vent shall be listed and installed in accordance with the terms of the special gas vent listing and the manufacturers' instructions.

503.5 Masonry, metal, and factory-built chimneys. Masonry, metal and factory-built chimneys shall comply with Sections 503.5.1 through 503.5.10.

503.5.1 Factory-built chimneys. Factory-built chimneys shall be installed in accordance with their listing and the manufacturers' instructions. Factory-built chimneys used to vent appliances that operate at positive vent pressure shall be listed for such application.

503.5.2 Metal chimneys. Metal chimneys shall be built and installed in accordance with NFPA 211, or local building codes.

503.5.3 Masonry chimneys. Masonry chimneys shall be built and installed in accordance with NFPA 211, or local building codes and shall be lined with approved clay flue lining, a listed chimney lining system, or other approved material that will resist corrosion, erosion, softening, or cracking from vent gases at temperatures up to 1800°F (982°C).

> **Exception:** Masonry chimney flues serving listed gas appliances with draft hoods, Category I appliances, and other gas appliances listed for use with Type B vent shall be permitted to be lined with a chimney lining system specifically listed for use only with such appliances. The liner shall be installed in accordance with the liner manufacturers' instructions and the terms of the listing. A permanent identifying label shall be attached at the point where the connection is to be made to the liner. The label shall read: "This chimney liner is for appliances that burn gas only. Do not connect to solid or liquid fuel-burning appliances or incinerators."
>
> For information on installation of gas vents in existing masonry chimneys, see Section 503.6.6.

503.5.4 Chimney termination. Chimneys for residential-type or low-heat gas utilization equipment shall extend at least 3 feet (914 mm) above the highest point where it passes through a roof of a building and at least 2 feet (610 mm) higher than any portion of a building within a horizontal distance of 10 feet (3048 mm) (see Figure 503.5.4). Chimneys for medium-heat equipment shall extend at least 10 feet (3048 mm) higher than any portion of any building within 25 feet (7620 mm). Chimneys shall extend at least 5 feet (1524 mm) above the highest connected equipment draft hood outlet or flue collar. Decorative shrouds shall not be installed at the termination of factory-built chimneys except where such shrouds are listed and labeled for use with the specific factory-built chimney system and are installed in accordance with the manufacturers' installation instructions.

FIGURE 503.5.4 − 503.5.6.3

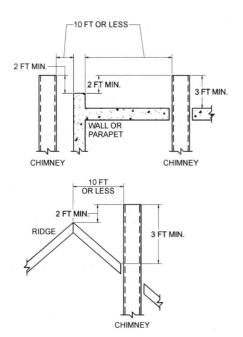

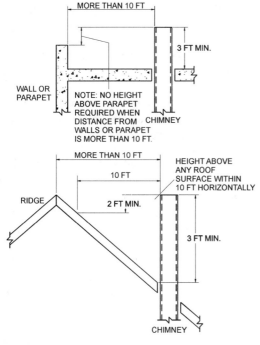

A. TERMINATION 10 FT OR LESS FROM RIDGE, WALL, OR PARAPET

B. TERMINATION MORE THAN 10 FT FROM RIDGE, WALL, OR PARAPET

For SI: 1 inch 25.4 mm, 1 foot = 304.8 mm.

FIGURE 503.5.4
TYPICAL TERMINATION LOCATIONS FOR
CHIMNEYS AND SINGLE-WALL METAL PIPES SERVING
RESIDENTIAL-TYPE AND LOW-HEAT EQUIPMENT

503.5.5 Size of chimneys. The effective area of a chimney venting system serving listed appliances with draft hoods, Category I appliances, and other appliances listed for use with Type B vents shall be in accordance with Section 504.

Where an incinerator is vented by a chimney serving other gas utilization equipment, the gas input to the incinerator shall not be included in calculating chimney size, provided the chimney flue diameter is not less than 1 inch (25 mm) larger in equivalent diameter than the diameter of the incinerator flue outlet.

Exceptions:

1. As an alternate method of sizing an individual chimney venting system for a single appliance with a draft hood, the effective areas of the vent connector and chimney flue shall be not less than the area of the appliance flue collar or draft hood outlet, nor greater than seven times the draft hood outlet area.

2. As an alternate method for sizing a chimney venting system connected to two appliances with draft hoods, the effective area of the chimney flue shall be not less than the area of the larger draft hood outlet plus 50 percent of the area of the smaller draft hood outlet, nor greater than seven times the smallest draft hood outlet area.

503.5.6 Inspection of chimneys. Before replacing an existing appliance or connecting a vent connector to a chimney, the chimney passageway shall be examined to ascertain that it is clear and free of obstructions and it shall be cleaned if previously used for venting solid or liquid fuel-burning appliances or fireplaces.

Exception: Existing chimneys shall be permitted to have their use continued when an appliance is replaced by an appliance of similar type, input rating, and efficiency.

503.5.6.1 Chimney lining. Chimneys shall be lined in accordance with NFPA 211.

503.5.6.2 Cleanouts. Cleanouts shall be examined to determine if they will remain tightly closed when not in use.

503.5.6.3 Unsafe chimneys. Where inspection reveals that an existing chimney is not safe for the intended application, it shall be repaired, rebuilt, lined, relined, or replaced with a vent or chimney to conform to NFPA 211, or local building codes, and it shall be suitable for the equipment to be vented.

503.5.7 Chimney serving equipment burning other fuels. Chimneys serving equipment burning other fuels shall comply with Sections 503.5.7.1 through 503.5.7.4.

503.5.7.1 Solid fuel-burning appliances. Gas utilization equipment shall not be connected to a chimney flue serving a separate appliance designed to burn solid fuel.

503.5.7.2 Liquid fuel-burning appliances. Gas utilization equipment and equipment burning liquid fuel shall be permitted to be connected to one chimney flue through separate openings or shall be permitted to be connected through a single opening if joined by a suitable fitting located as close as practical to the chimney. If two or more openings are provided into one chimney flue, they shall be at different levels. If the gas utilization equipment is automatically controlled, it shall be equipped with a safety shutoff device.

503.5.7.3 Combination gas and solid fuel-burning appliances. A combination gas- and solid fuel-burning appliance equipped with a manual reset device to shut off gas to the main burner in the event of sustained backdraft or flue gas spillage shall be permitted to be connected to a single chimney flue. The chimney flue shall be sized to properly vent the appliance.

503.5.7.4 Combination gas and oil fuel-burning appliances. A listed combination gas- and oil-burning appliance shall be permitted to be connected to a single chimney flue. The chimney flue shall be sized to properly vent the appliance.

503.5.8 Support of chimneys. All portions of chimneys shall be supported for the design and weight of the materials employed. Factory-built chimneys shall be supported and spaced in accordance with their listings and the manufacturer's instructions.

503.5.9 Cleanouts. Where a chimney that formerly carried flue products from liquid or solid fuel-burning appliances is used with an appliance using fuel gas, an accessible cleanout shall be provided. The cleanout shall have a tight-fitting cover and shall be installed so its upper edge is at least 6 inches (152 mm) below the lower edge of the lowest chimney inlet opening.

503.5.10 Space surrounding lining or vent. The remaining space surrounding a chimney liner, gas vent, special gas vent, or plastic piping installed within a masonry chimney flue shall not be used to vent another appliance.

Exception: The insertion of another liner or vent within the chimney as provided in this code and the liner or vent manufacturer's instructions.

503.6 Gas vents. Gas vents shall comply with sections 503.6.1 through 503.6.12 (See Section 202, Definitions).

503.6.1 Installation, general. Gas vents shall be installed in accordance with the terms of their listings and the manufacturer's instructions.

503.6.2 Type B-W vent capacity. A Type B-W gas vent shall have a listed capacity not less than that of the listed vented wall furnace to which it is connected.

503.6.3 Roof penetration. A gas vent passing through a roof shall extend through the roof flashing, roof jack, or roof thimble and shall be terminated by a listed termination cap.

503.6.4 Offsets. Type B and Type L vents shall extend in a generally vertical direction with offsets not exceeding 45 degrees (0.79 rad), except that a vent system having not more than one 60-degree (1.04 rad) offset shall be permitted. Any angle greater than 45 degrees (0.79 rad) from the vertical is considered horizontal. The total horizontal length of a vent plus the horizontal vent connector length serving draft hood-equipped appliances shall not be greater than 75 percent of the vertical height of the vent.

Exception: Systems designed and sized as provided in Section 504. ⇐

Vents serving Category I fan-assisted appliances shall be installed in accordance with the appliance manufacturer's instructions and Section 504. ⇐

503.6.5 Gas vents installed within masonry chimneys. Gas vents installed within masonry chimneys shall be installed in accordance with the terms of their listing and the manufacturer's installation instructions. Gas vents installed within masonry chimneys shall be identified with a permanent label installed at the point where the vent enters the chimney. The label shall contain the following language: "This gas vent is for appliances that burn gas. Do not connect to solid or liquid fuel-burning appliances or incinerators."

503.6.6 Gas vent terminations. A gas vent shall terminate above the roof surface with a listed cap or listed roof assembly. Gas vents 12 inches (305 mm) in size or smaller with listed caps shall be permitted to be terminated in accordance with Figure 503.6.6, provided that such vents are at least 8 feet (2438 mm) from a vertical wall or similar obstruction. All other gas vents shall terminate not less than 2 feet (610 mm) above the highest point where they pass through the roof and at least 2 feet (610 mm) higher than any portion of a building within 10 feet (3048 mm).

Exceptions:

1. Industrial equipment as provided in Section 503.2.2.
2. Direct vent systems as provided in Section 503.2.3.
3. Equipment with integral vents as provided in Section 503.2.4.
4. Mechanical draft systems as provided in Section 503.3.3.
5. Ventilating hoods and exhaust systems as provided in Section 503.3.4.

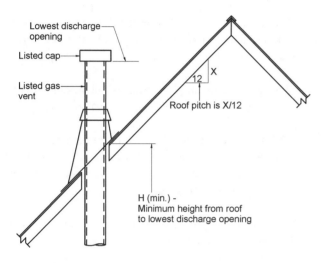

503.6.7 Minimum height. A Type B or a Type L gas vent shall terminate at least 5 feet (1524 mm) in vertical height above the highest connected equipment draft hood or flue collar. A Type B-W gas vent shall terminate at least 12 feet (3658 mm) in vertical height above the bottom of the wall furnace.

503.6.8 Exterior wall penetrations. A gas vent extending through an exterior wall shall not terminate adjacent to the wall or below eaves or parapets, except as provided in Sections 503.2.3 and 503.3.3.

503.6.9 Size of gas vents. Venting systems shall be sized and constructed in accordance with Section 504 and the gas vent and gas equipment manufacturer's instructions.

503.6.9.1 Category I appliances. The sizing of natural draft venting systems serving one or more listed appliances equipped with a draft hood or appliances listed for use with Type B gas vent, installed in a single story of a building, shall be in accordance with Section 504 or in accordance with sound engineering practice. Category I appliances are either draft hood equipped or fan-assisted combustion system in design. Different vent design methods are required for draft hood-equipped and fan-assisted combustion system appliances.

Exceptions:

1. As an alternate method for sizing an individual gas vent for a single, draft hood-equipped appliance, the effective area of the vent connector and the gas vent shall be not less than the area of the appliance draft hood outlet, nor greater than seven times the draft hood outlet area. Vents serving fan-assisted combustion system appliances shall be sized in accordance with Section 504. ⇐

2. As an alternate method for sizing a gas vent connected to two appliances with draft hoods, the effective area of the vent shall be not less than the area of the larger draft hood outlet plus 50 percent of the smaller draft hood outlets, nor greater than seven times the smallest draft hood outlet area. Vents serving fan-assisted combustion system appliances, or combinations of fan-assisted combustion system and draft hood-equipped appliances, shall be sized in accordance with Section 504. ⇐

503.6.9.2 Category II, III, and IV appliances. The sizing of gas vents for Category II, III, and IV equipment shall be in accordance with the equipment manufacturer's instructions.

ROOF PITCH	H (min) ft
Flat to 6/12	1.0
6/12 to 7/12	1.25
Over 7/12 to 8/12	1.5
Over 8/12 to 9/12	2.0
Over 9/12 to 10/12	2.5
Over 10/12 to 11/12	3.25
Over 11/12 to 12/12	4.0
Over 12/12 to 14/12	5.0
Over 14/12 to 16/12	6.0
Over 16/12 to 18/12	7.0
Over 18/12 to 20/12	7.5
Over 20/12 to 21/12	8.0

For SI: 1 inch = 25.4 mm, 1 foot = 304.8 mm.

FIGURE 503.6.6
GAS VENT TERMINATION LOCATIONS FOR
LISTED CAPS 12 INCHES OR LESS IN SIZE AT
LEAST 8 FEET FROM A VERTICAL WALL

503.6.10 Gas vents serving equipment on more than one floor. A single or common gas vent shall be permitted in multistory installations to vent Category I equipment located on more than one floor level, provided the venting system is designed and installed in accordance with this section and approved engineering methods.

503.6.10.1 Equipment separation. All equipment connected to the common vent shall be located in rooms separated from habitable space. Each of these rooms shall have provisions for an adequate supply of combustion, ventilation, and dilution air that is not supplied from habitable space (see Figure 503.6.10.1).

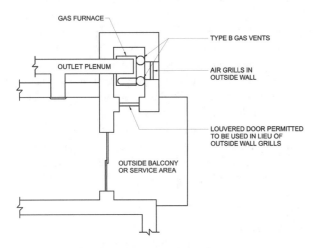

FIGURE 503.6.10.1
PLAN VIEW OF PRACTICAL SEPARATION METHOD
FOR MULTISTORY GAS VENTING

503.6.10.2 Sizing. The size of the connectors and common segments of multistory venting systems for equipment listed for use with Type B double-wall gas vent shall be in accordance with Table 504.3(1) and Figures B-13 and B-14 in Appendix B, provided:

1. The available total height *(H)* for each segment of a multistory venting system is the vertical distance between the level of the highest draft hood outlet or flue collar on that floor and the centerline of the next highest interconnection tee. (See Figure B-13).
2. The size of the connector for a segment is determined from its gas utilization equipment heat input and available connector rise, and shall not be smaller than the draft hood outlet or flue collar size.
3. The size of the common vertical segment, and of the interconnection tee at the base of that segment, shall be based on the total gas utilization equipment heat input entering that segment and its available total height.

503.6.11 Support of gas vents. Gas vents shall be supported and spaced in accordance with their listings and the manufacturers' instructions.

503.6.12 Marking. In those localities where solid and liquid fuels are used extensively, gas vents shall be permanently identified by a label attached to the wall or ceiling at a point where the vent connector enters the gas vent. The label shall read:

"This gas vent is for appliances that burn gas. Do not connect to solid or liquid fuel-burning appliances or incinerators."

503.7 Single-wall metal pipe. Single-wall metal pipe vents shall comply with Sections 503.7.1 through 503.7.12.

503.7.1 Construction. Single-wall metal pipe shall be constructed of galvanized sheet steel not less than 0.0304 inch (0.7 mm) thick, or other approved, noncombustible, corrosion-resistant material.

503.7.2 Cold climate. Uninsulated single-wall metal pipe shall not be used outdoors in cold climates for venting gas utilization equipment.

503.7.3 Termination. Single-wall metal pipe shall terminate at least 5 feet (1524 mm) in vertical height above the highest connected equipment draft hood outlet or flue collar. Single-wall metal pipe shall extend at least 2 feet (610 mm) above the highest point where it passes through a roof of a building and at least 2 feet (610 mm) higher than any portion of a building within a horizontal distance of 10 feet (3048 mm) (See Figure 503.5.4). An approved cap or roof assembly shall be attached to the terminus of a single-wall metal pipe [see also Section 503.7.8, item 3].

503.7.4 Limitations of use. Single-wall metal pipe shall be used only for runs directly from the space in which the equipment is located through the roof or exterior wall to the outdoor atmosphere.

503.7.5 Roof penetrations. A pipe passing through a roof shall extend without interruption through the roof flashing, roof jacket, or roof thimble. Where a single-wall metal pipe passes through a roof constructed of combustible material, a noncombustible, nonventilating thimble shall be used at the point of passage. The thimble shall extend at least 18 inches (457 mm) above and 6 inches (152 mm) below the roof with the annular space open at the bottom and closed only at the top. The thimble shall be sized in accordance with Section 503.10.16.

TABLE 503.7.7
CLEARANCES FOR CONNECTORS[a]

EQUIPMENT	MINIMUM DISTANCE FROM COMBUSTIBLE MATERIAL			
	Listed Type B gas vent material	Listed Type L vent material	Single-wall metal pipe	Factory-built chimney sections
Listed equipment with draft hoods and equipment listed for use with Type B gas vents	As listed	As listed	6 inches	As listed
Residential boilers and furnaces with listed gas conversion burner and with draft hood	6 inches	6 inches	9 inches	As listed
Residential appliances listed for use with Type L vents	Not permitted	As listed	9 inches	As listed
Residential incinerators	Not permitted	9 inches	18 inches	As listed
Listed gas-fired toilets	Not permitted	As listed	As listed	As listed
Unlisted residential appliances with draft hood	Not permitted	6 inches	9 inches	As listed
Residential and low-heat equipment other than above	Not permitted	9 inches	18 inches	As listed
Medium-heat equipment	Not permitted	Not permitted	36 inches	As listed

For SI: 1 inch = 25.4 mm.

a. These clearances shall apply unless the listing of an appliance or connector specifies different clearances, in which case the listed clearances shall apply.

503.7.6 Installation. Single-wall metal pipe shall not originate in any unoccupied attic or concealed space and shall not pass through any attic, inside wall, concealed space, or floor. The installation of a single-wall metal pipe through an exterior combustible wall shall comply with Section 503.10.16. Single-wall metal pipe used for venting an incinerator shall be exposed and readily examinable for its full length and shall have suitable clearances maintained.

503.7.7 Clearances. Minimum clearances from single-wall metal pipe to combustible material shall be in accordance with Table 503.7.7. The clearance from single-wall metal pipe to combustible material shall be permitted to be reduced where the combustible material is protected as specified for vent connectors in Table 308.2.

503.7.8 Size of single-wall metal pipe. A venting system constructed of single-wall metal pipe shall be sized in accordance with one of the following methods and the equipment manufacturer's instructions:

1. For a draft-hood-equipped appliance, in accordance with Section 504.
2. For a venting system for a single appliance with a draft hood, the areas of the connector and the pipe each shall be not less than the area of the appliance flue collar or draft hood outlet, whichever is smaller. The vent area shall not be greater than seven times the draft hood outlet area.

3. Other approved engineering methods.

503.7.9 Pipe geometry. Any shaped single-wall metal pipe shall be permitted to be used, provided that its equivalent effective area is equal to the effective area of the round pipe for which it is substituted, and provided that the minimum internal dimension of the pipe is not less than 2 inches (51 mm).

503.7.10 Termination capacity. The vent cap or a roof assembly shall have a venting capacity not less than that of the pipe to which it is attached.

503.7.11 Support of single-wall metal pipe. All portions of single-wall metal pipe shall be supported for the design and weight of the material employed.

503.7.12 Marking. Single-wall metal pipe shall comply with the marking provisions of Section 503.6.12.

503.8 Venting system termination location. The location of venting system terminations shall comply with the following and Figure 503.8:

1. A mechanical draft venting system shall terminate at least 3 feet (914 mm) above any forced-air inlet located within 10 feet (3048 mm).

FIGURE 503.8 – 503.10.2.1   CHIMNEYS AND VENTS

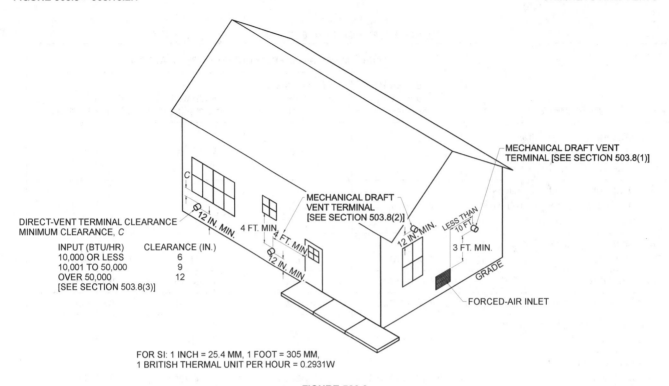

DIRECT-VENT TERMINAL CLEARANCE
MINIMUM CLEARANCE, *C*

INPUT (BTU/HR)	CLEARANCE (IN.)
10,000 OR LESS	6
10,001 TO 50,000	9
OVER 50,000	12
[SEE SECTION 503.8(3)]	

FOR SI: 1 INCH = 25.4 MM, 1 FOOT = 305 MM,
1 BRITISH THERMAL UNIT PER HOUR = 0.2931W

FIGURE 503.8
EXIT TERMINALS OF MECHANICAL DRAFT AND DIRECT-VENT SYSTEM

Exceptions:

1. This provision shall not apply to the combustion air intake of a direct-vent appliance.

2. This provision shall not apply to the separation of the integral outdoor air inlet and flue gas discharge of listed outdoor appliances.

2. A mechanical draft venting system, excluding direct-vent appliances, shall terminate at least 4 feet (1219 mm) below, 4 feet (1219 mm) horizontally from, or 1 foot (305 mm) above any door, window, or gravity air inlet into any building. The bottom of the vent terminal shall be located at least 12 inches (305 mm) above grade.

3. The vent terminal of a direct-vent appliance with an input of 10,000 Btu per hour (3 kW) or less shall be located at least 6 inches (152 mm) from any air opening into a building; such an appliance with an input over 10,000 Btu per hour (3 kW) but not over 50,000 Btu per hour (14.7 kW) shall be installed with a 9-inch (230 mm) vent termination clearance; and an appliance with an input over 50,000 Btu per hour (14.7 kW) shall have at least a 12-inch (305 mm) vent termination clearance. The bottom of the vent terminal and the air intake shall be located at least 12 inches (305 mm) above grade.

4. Through-the-wall vents for Category II and IV appliances and noncategorized condensing appliances shall not terminate over public walkways or over an area where condensate or vapor could create a nuisance or hazard or could be detrimental to the operation of regulators, relief valves, or other equipment.

503.9 Condensation drainage. Provision shall be made to collect and dispose of condensate from venting systems serving Category II and IV equipment and noncategorized condensing appliances in accordance with Section 503.8(4). Where local experience indicates that condensation is a problem, provision shall be made to drain off and dispose of condensate from venting systems serving Category I and III equipment in accordance with Section 503.8(4).

503.10 Vent connectors for Category I equipment. Vent connectors for Category I equipment shall comply with Sections 503.10.1 through 503.10.17.

503.10.1 Where required. A vent connector shall be used to connect equipment to a gas vent, chimney, or single-wall metal pipe, except where the gas vent, chimney, or single-wall metal pipe is directly connected to the equipment.

503.10.2 Materials. Vent connectors shall be constructed in accordance with Sections 503.10.2.1 through 503.10.2.5.

503.10.2.1 General. A vent connector shall be made of noncombustible corrosion-resistant material capable of withstanding the vent gas temperature produced by the equipment and of sufficient thickness to withstand physical damage.

503.10.2.2 Vent connectors located in unconditioned areas. Where the vent connector used for equipment having a draft hood or a Category I appliance is located in or passes through an attic space or other unconditioned area, that portion of the vent connector shall be listed Type B or Type L or listed vent material or listed material having equivalent insulation qualities.

503.10.2.3 Residential-type appliance connectors. Where vent connectors for residential-type appliances are not installed in attics or other unconditioned spaces, connectors for listed appliances having draft hoods and for appliances having draft hoods and equipped with listed conversion burners shall be one of the following:

1. Type B or Type L vent material;
2. Galvanized sheet steel not less than 0.018 inch (0.46 mm) thick;
3. Aluminum (1100 or 3003 alloy or equivalent) sheet not less than 0.027 inch (0.69 mm) thick;
4. Stainless steel sheet not less than 0.012 inch (0.31 mm) thick;
5. Smooth interior wall metal pipe having resistance to heat and corrosion equal to or greater than that of Item 2, 3 or 4 above; or
6. A listed vent connector.

Vent connectors shall not be covered with insulation.

Exception: Listed insulated vent connectors shall be installed according to the terms of their listing.

503.10.2.4 Low-heat equipment. A vent connector for low-heat equipment shall be a factory-built chimney section or steel pipe having resistance to heat and corrosion equivalent to that for the appropriate galvanized pipe as specified in Table 503.10.2.4. Factory-built chimney sections shall be joined together in accordance with the chimney manufacturers' instructions.

503.10.2.5 Medium-heat appliances. Vent connectors for medium-heat equipment and commercial and industrial incinerators shall be constructed of factory-built medium-heat chimney sections or steel of a thickness not less than that specified in Table 503.10.2.5 and shall comply with the following:

1. A steel vent connector for equipment with a vent gas temperature in excess of 1000°F (538°C), measured at the entrance to the connector, shall be lined with medium-duty fire brick (ASTM C 64, Type F), or the equivalent.
2. The lining shall be at least $2^1/_2$ inches (64 mm) thick for a vent connector having a diameter or greatest cross-sectional dimension of 18 inches (457 mm) or less.
3. The lining shall be at least $4^1/_2$ inches (114 mm) thick laid on the $4^1/_2$-inch (114 mm) bed for a vent connector having a diameter or greatest cross-sectional dimension greater than 18 inches (457 mm).

TABLE 503.10.2.4
MINIMUM THICKNESS FOR GALVANIZED STEEL VENT CONNECTORS FOR LOW-HEAT APPLIANCES

DIAMETER OF CONNECTOR (inches)	MINIMUM THICKNESS (inch)
Less than 6	0.019
6 to less than 10	0.023
10 to 12 inclusive	0.029
14 to 16 inclusive	0.034
Over 16	0.056

For SI: 1 inch = 2.54 mm.

TABLE 503.10.2.5
MINIMUM THICKNESS FOR STEEL VENT CONNECTORS FOR MEDIUM-HEAT EQUIPMENT AND COMMERCIAL AND INDUSTRIAL INCINERATORS VENT CONNECTOR SIZE

DIAMETER (inches)	AREA (square inches)	MINIMUM THICKNESS (inch)
Up to 14	Up to 154	0.053
Over 14 to 16	154 to 201	0.067
Over 16 to 18	201 to 254	0.093
Over 18	Larger than 254	0.123

For SI: 1 inch = 25.4 mm, 1 square inch = 645.16 mm^2.

4. Factory-built chimney sections, if employed, shall be joined together in accordance with the chimney manufacturer's instructions.

503.10.3 Size of vent connector. Vent connectors shall be sized in accordance with Sections 503.10.3.1 through 503.10.3.6.

503.10.3.1 Single draft hood and fan-assisted. A vent connector for equipment with a single draft hood or for a Category I fan-assisted combustion system appliance shall be sized and constructed in accordance with Section 504.

503.10.3.2 Multiple draft hood. For a single appliance having more than one draft hood outlet or flue collar, the manifold shall be constructed according to the instructions of the appliance manufacturer. If there are no instructions, the manifold shall be designed and constructed in accordance with approved engineering practices. As an alternate method, the effective area of the manifold shall equal the combined area of the flue collars or draft hood outlets and the vent connectors shall have a minimum 1-foot (305 mm) rise.

503.10.3.3 Multiple appliances. Where two or more appliances are connected to a common vent or chimney, each vent connector shall be sized in accordance with Section 504.

As an alternative method applicable only when all of the appliances are draft hood equipped, each vent connector shall have an effective area not less than the area of the draft hood outlet of the appliance to which it is connected.

503.10.3.4 Common connector/manifold. Where two or more gas appliances are vented through a common vent connector or vent manifold, the common vent connector or vent manifold shall be located at the highest level consistent with available headroom and the required clearance to combustible materials and shall be sized in accordance with Section 504.

As an alternate method applicable only where there are two draft hood-equipped appliances, the effective area of the common vent connector or vent manifold and all junction fittings shall be not less than the area of the larger vent connector plus 50 percent of the areas of the smaller flue collar outlet.

503.10.3.5 Size increase. Where the size of a vent connector is increased to overcome installation limitations and obtain connector capacity equal to the equipment input, the size increase shall be made at the equipment draft hood outlet.

503.10.3.6 Approved engineering practices. The effective area of the vent connector, where connected to one or more appliances requiring draft for operation, shall be obtained by the application of approved engineering practices to perform as specified in Sections 503.3 and 503.3.1.

503.10.4 Two or more appliances connected to a single vent. Where two or more vent connectors enter a common gas vent, chimney flue, or single-wall metal pipe, the smaller connector shall enter at the highest level consistent with the available headroom or clearance to combustible material. Vent connectors serving Category I appliances shall not be connected to any portion of a mechanical draft system operating under positive static pressure, such as those serving Category III or IV appliances.

503.10.5 Clearance. Minimum clearances from vent connectors to combustible material shall be in accordance with Table 503.7.7.

Exception: The clearance between a vent connector and combustible material shall be permitted to be reduced where the combustible material is protected as specified for vent connectors in Table 308.2.

503.10.6 Flow resistance. A vent connector shall be installed so as to avoid turns or other construction features that create excessive resistance to flow of vent gases.

503.10.7 Joints. Joints between sections of connector piping and connections to flue collars and hood outlets shall be fastened by sheet-metal screws or other approved means.

Exception: Vent connectors of listed vent material, assembled and connected to flue collars and draft hood outlets in accordance with the manufacturers' instructions.

503.10.8 Slope. A vent connector shall be installed without dips or sags and shall slope upward toward the vent or chimney at least $^1/_4$ inch per foot (21 mm/m).

503.10.9 Length of vent connector. A vent connector shall be as short as practical and the equipment located as close as practical to the chimney or vent. Except as provided for in Section 503.10.3, the maximum horizontal length of a single-wall connector shall be 75 percent of the height of the chimney or vent. Except as provided for in Section 503.10.3, the maximum horizontal length of a Type B double-wall connector shall be 100 percent of the height of the chimney or vent. For a chimney or vent system serving multiple appliances, the maximum length of an individual connector, from the appliance outlet to the junction with the common vent or another connector, shall be 100 percent of the height of the chimney or vent.

503.10.10 Support. A vent connector shall be supported for the design and weight of the material employed to maintain clearances and prevent physical damage and separation of joints.

503.10.11 Location. Where the vent connector used for equipment having a draft hood or for Category I appliances is located in or passes through an attic, crawl space, or other unconditioned area subject to low ambient temperatures, that portion of the vent connector shall be of listed double-wall Type B, Type L vent material or listed material having equivalent insulation qualities.

503.10.12 Chimney connection. Where entering a flue in a masonry or metal chimney, the vent connector shall be installed above the extreme bottom to avoid stoppage. A thimble or slip joint shall be permitted to be used to facilitate removal of the connector. The connector shall be firmly attached to or inserted into the thimble or slip joint to prevent the connector from falling out. Means shall be employed to prevent the connector from entering so far as to restrict the space between its end and the opposite wall of the chimney flue (see Section 501.9).

503.10.13 Inspection. The entire length of a vent connector shall be provided with ready access for inspection, cleaning, and replacement.

503.10.14 Fireplaces. A vent connector shall not be connected to a chimney flue serving a fireplace unless the fireplace flue opening is permanently sealed.

503.10.15 Passage through ceilings, floors, or walls. A vent connector shall not pass through any ceiling, floor or fire-resistance rated wall. A single-wall metal pipe connector shall not pass through any interior wall.

> **Exception:** Vent connectors made of listed Type B or Type L vent material and serving listed equipment with draft hoods and other equipment listed for use with Type B gas vents shall be permitted to pass through walls or partitions constructed of combustible material if the connectors are installed with not less than the listed clearance to combustible material.

503.10.16 Single-wall connector penetrations of combustible walls. A vent connector made of a single-wall metal pipe shall not pass through a combustible exterior wall unless guarded at the point of passage by a ventilated metal thimble not smaller than the following:

1. For listed equipment equipped with draft hoods and equipment listed for use with Type B gas vents, the thimble shall be not less than 4 inches (102 mm) larger in diameter than the vent connector. Where there is a run of not less than 6 feet (1829 mm) of vent connector in the open between the draft hood outlet and the thimble, the thimble shall be permitted to be not less than 2 inches (51 mm) larger in diameter than the vent connector.
2. For unlisted equipment having draft hoods, the thimble shall be not less than 6 inches (152 mm) larger in diameter than the vent connector.
3. For residential incinerators and all other residential and low-heat equipment, the thimble shall be not less than 12 inches (305 mm) larger in diameter than the vent connector.

> **Exception:** In lieu of thimble protection, all combustible material in the wall shall be removed from the vent connector a sufficient distance to provide the specified clearance from such vent connector to combustible material. Any material used to close up such opening shall be noncombustible.

503.10.17 Medium-heat connectors. Vent connectors for medium-heat equipment shall not pass through walls or partitions constructed of combustible material.

503.11 Vent connectors for Category II, III, and IV gas utilization equipment. Vent connectors for Category II, III and IV appliances shall be as specified for the venting systems in accordance with Section 503.4.

503.12 Draft hoods and draft controls. The installation of draft hoods and draft controls shall comply with Sections 503.12.1 through 503.12.8.

503.12.1 Equipment requiring draft hoods. Vented equipment shall be installed with draft hoods.

> **Exception:** Dual oven type combination ranges, incinerators, direct-vent equipment, fan-assisted combustion system appliances, equipment requiring chimney draft for operation, single firebox boilers equipped with conversion burners with inputs greater than 400,000 Btu per hour (117 Kw), equipment equipped with blast, power, or pressure burners that are not listed for use with draft hoods, and equipment designed for forced venting.

503.12.2 Installation. A draft hood supplied with or forming a part of listed vented equipment shall be installed without alteration, exactly as furnished and specified by the equipment manufacturer. If a draft hood is not supplied by the equipment manufacturer where one is required, a draft hood shall be installed, shall be of a listed or approved type and, in the absence of other instructions, shall be of the same size as the equipment flue collar. Where a draft hood is required with a conversion burner, it shall be of a listed or approved type.

> **Exception:** Where it is determined that a draft hood of special design is needed or preferable for a particular installation, the installation shall be in accordance with the recommendations of the equipment manufacturer and shall be approved.

503.12.3 Draft control devices. Where a draft control device is part of the equipment or is supplied by the equipment manufacturer, it shall be installed in accordance with the manufacturer's instructions. In the absence of manufacturers' instructions, the device shall be attached to the flue collar of the equipment or as near to the equipment as practical.

503.12.4 Additional devices. Equipment (except incinerators) requiring controlled chimney draft shall be permitted to be equipped with a listed double-acting barometric-draft regulator installed and adjusted in accordance with the manufacturer's instructions.

503.12.5 Incinerator draft regulator. A listed incinerator shall be permitted to be equipped with a listed single-acting barometric draft regulator where recommended by the incinerator manufacturer. This draft regulator shall be installed in accordance with the incinerator manufacturer's instructions.

503.12.6 Location. Draft hoods and barometric draft regulators shall be installed in the same room or enclosure as the equipment in such a manner as to prevent any difference in pressure between the hood or regulator and the combustion air supply.

503.12.7 Positioning. Draft hoods and draft regulators shall be installed in the position for which they were designed with reference to the horizontal and vertical planes and shall be located so that the relief opening is not obstructed by any part of the equipment or adjacent construction. The equipment and its draft hood shall be located so that the relief opening is accessible for checking vent operation.

503.12.8 Clearance. A draft hood shall be located so its relief opening is not less than 6 inches (152 mm) from any surface except that of the equipment it serves and the venting system to which the draft hood is connected. Where a greater or lesser clearance is indicated on the equipment label, the clearance shall be not less than that specified on the label. Such clearances shall not be reduced.

503.13 Manually operated dampers. A manually operated damper shall not be placed in the vent connector for any equipment, except in a connector serving a listed incinerator where recommended by the incinerator manufacturer and installed in accordance with the incinerator manufacturer's instructions. Fixed baffles shall not be classified as manually operated dampers.

503.14 Automatically operated vent dampers. An automatically operated vent damper shall be of a listed type.

503.15 Obstructions. A device that retards the flow of vent gases shall not be installed in a vent connector, chimney, or vent. The tables in Section 504 shall not apply where the devices covered in this section are installed in the vent.

Exceptions:
1. Draft regulators and safety controls specifically listed for installation in venting systems and installed in accordance with the terms of their listing.
2. Draft regulators and safety controls that are designed and installed in accordance with approved engineering methods and that are approved.
3. Listed heat reclaimers and automatically operated vent dampers installed in accordance with the terms of their listing.
4. Approved economizers, heat reclaimers, and recuperators installed in venting systems of equipment not required to be equipped with draft hoods, provided the gas utilization equipment manufacturer's instructions cover the installation of such a device in the venting system and performance in accordance with Sections 503.3 and 503.3.1 is obtained.

SECTION 504
SIZING OF CATEGORY I APPLIANCE
VENTING SYSTEMS

504.1 Definitions. The following definitions apply to the tables in this section.

APPLIANCE CATEGORIZED VENT DIAMETER/AREA. The minimum vent area/diameter permissible for Category I appliances to maintain a nonpositive vent static pressure when tested in accordance with nationally recognized standards.

FAN-ASSISTED COMBUSTION SYSTEM. An appliance equipped with an integral mechanical means to either draw or force products of combustion through the combustion chamber or heat exchanger.

FAN Min. The minimum input rating of a Category I fan-assisted appliance attached to a vent or connector.

FAN Max. The maximum input rating of a Category I fan-assisted appliance attached to a vent or connector.

NAT Max. The maximum input rating of a Category I draft-hood-equipped appliance attached to a vent or connector.

FAN + FAN. The maximum combined appliance input rating of two or more Category I fan-assisted appliances attached to the common vent.

FAN + NAT. The maximum combined appliance input rating of one or more Category I fan-assisted appliances and one or more Category I draft-hood-equipped appliances attached to the common vent.

NA. Vent configuration is not allowed due to potential for condensate formation or pressurization of the venting system, or not applicable due to physical or geometric restraints.

NAT + NAT. The maximum combined appliance input rating of two or more Category I draft-hood-equipped appliances attached to the common vent.

504.2 Application of single-appliance vent Tables 504.2(1) through 504.2(5). The application of Tables 504.2(1) through 504.2(5) shall be subject to the requirements of Sections 504.2.1 through 504.2.13.

504.2.1 Vent obstructions. These venting tables shall not be used where obstructions, as described in the exceptions to Section 503.15, are installed in the venting system. The installation of vents serving listed appliances with vent dampers shall be in accordance with the appliance manufacturer's instructions or in accordance with the following:
1. The maximum capacity of the vent system shall be determined using the "NAT Max" column.
2. The minimum capacity shall be determined as if the appliance were a fan-assisted appliance, using the "FAN Min" column to determine the minimum capacity of the vent system. Where the corresponding "FAN Min" is "NA," the vent configuration shall not be permitted and an alternative venting configuration shall be utilized.

504.2.2 Minimum size. Where the vent size determined from the tables is smaller than the appliance draft hood outlet or flue collar, the smaller size shall be permitted to be used provided all of the following requirements are met:

1. The total vent height (H) is at least 10 feet (3048 mm).
2. Vents for appliance draft hood outlets or flue collars 12 inches (305 mm) in diameter or smaller are not reduced more than one table size.
3. Vents for appliance draft hood outlets or flue collars larger than 12 inches (305 mm) in diameter are not reduced more than two table sizes.
4. The maximum capacity listed in the tables for a fan-assisted appliance is reduced by 10 percent (0.90 x maximum table capacity).
5. The draft hood outlet is greater than 4 inches (102 mm) in diameter. Do not connect a 3-inch (76 mm) diameter vent to a 4-inch diameter (102 mm) draft hood outlet. This provision shall not apply to fan-assisted appliances.

504.2.3 Vent offsets. Single-appliance venting configurations with zero (0) lateral lengths in Tables 504.2(1), 504.2(2), and 504.2(5) shall not have elbows in the venting system. For vent configurations with lateral lengths, the venting tables include allowance for two 90-degree turns. For each additional 90-degree (1.6 rad) turn, or equivalent, the maximum capacity listed in the venting tables shall be reduced by 10 percent (0.90 x maximum table capacity). Two or more turns, the combined angles of which equal 90 degrees, shall be considered equivalent to one 90-degree (1.6 rad) turn.

504.2.4 Zero lateral. Zero (0) lateral (L) shall apply only to a straight vertical vent attached to a top outlet draft hood or flue collar.

504.2.5 High-altitude installations. Sea-level input ratings shall be used when determining maximum capacity for high altitude installation. Actual input (derated for altitude) shall be used for determining minimum capacity for high altitude installation.

504.2.6 Multiple input rate appliances. For appliances with more than one input rate, the minimum vent capacity (FAN Min) determined from the tables shall be less than the lowest appliance input rating, and the maximum vent capacity (FAN Max/NAT Max) determined from the tables shall be greater than the highest appliance rating input.

504.2.7 Liner system sizing. Listed corrugated metallic chimney liner systems in masonry chimneys shall be sized by using Table 504.2(1) or 504.2(2) for Type B vents with the maximum capacity reduced by 20 percent (0.80 x maximum capacity) and the minimum capacity as shown in Table 504.2(1) or 504.2(2). Corrugated metallic liner systems installed with bends or offsets shall have their maximum capacity further reduced in accordance with Section 504.2.3.

504.2.8 Vent area and diameter. Where the vertical vent has a larger diameter than the vent connector, the vertical vent diameter shall be used to determine the minimum vent capacity, and the connector diameter shall be used to determine the maximum vent capacity. The flow area of the vertical vent shall not exceed seven times the flow area of the listed appliance categorized vent area, flue collar area, or draft hood outlet area unless designated in accordance with approved engineering methods.

504.2.9 Chimney and vent locations. Tables 504.2(1), 504.2(2), 504.2(3), 504.2(4) and 504.2(5) shall be used for chimneys and vents not exposed to the outdoors below the roof line. A Type B vent or listed chimney lining system passing through an unused masonry chimney flue shall not be considered to be exposed to the outdoors. Table 504.2(3) in combination with Table 504.3(6) shall be used for clay-tile-lined exterior masonry chimneys, provided all of the following are met:

1. Vent connector is Type B double-wall.
2. Vent connector length is limited to $1^1/_2$ feet for each inch (18 mm per mm) of vent connector diameter.
3. The appliance is draft hood equipped.
4. The input rating is less than the maximum capacity given by Table 504.2(2).
5. For a water heater, the outdoor design temperature is not less than 5°F (−15°C).
6. For a space-heating appliance, the input rating is greater than the minimum capacity given by Table 504.3(6).

Where these conditions cannot be met, an alternative venting design shall be used, such as a listed chimney lining system.

Exception: The installation of vents serving listed appliances shall be permitted to be in accordance with the appliance manufacturer's instructions and the terms of the listing.

504.2.10 Vent connector size limitation. Vent connectors shall not be increased in size more than two sizes greater than the listed appliance categorized vent diameter, flue collar diameter, or draft hood outlet diameter.

504.2.11 Component commingling. In a single run of vent or vent connector, different diameters and types of vent and connector components shall be permitted to be used, provided that all such sizes and types are permitted by the tables.

504.2.12 Table interpolation. Interpolation shall be permitted in calculating capacities for vent dimensions that fall between the table entries (see Example 3, Appendix B).

504.2.13 Extrapolation prohibited. Extrapolation beyond the table entries shall not be permitted.

504.2.14 Engineering calculations. For vent heights less than 6 feet (1829 mm) and greater than shown in the tables, engineering methods shall be used to calculate vent capacities.

504.3 Application of multiple appliance vent Tables 504.3(1) through 504.3(8). The application of Tables 504.3(1) through 504.3(8) shall be subject to the requirements of Sections 504.3.1 through 504.3.23.

504.3.1 Vent obstructions. These venting tables shall not be used where obstructions, as described in the exceptions to Section 503.15, are installed in the venting system. The installation of vents serving listed appliances with vent dampers shall be in accordance with the appliance manufacturer's instructions or in accordance with the following:

1. The maximum capacity of the vent connector shall be determined using the NAT Max column.

2. The maximum capacity of the vertical vent or chimney shall be determined using the FAN+NAT column when the second appliance is a fan-assisted appliance, or the NAT+NAT column when the second appliance is equipped with a draft hood.

3. The minimum capacity shall be determined as if the appliance were a fan-assisted appliance.

    3.1 The minimum capacity of the vent connector shall be determined using the FAN Min column.

    3.2 The FAN+FAN column shall be used where the second appliance is a fan-assisted appliance, and the FAN+NAT column shall be used where the second appliance is equipped with a draft hood, to determine whether the vertical vent or chimney configuration is not permitted (NA). Where the vent configuration is NA, the vent configuration shall not be permitted and an alternative venting configuration shall be utilized.

504.3.2 Connector length limit. The vent connector shall be routed to the vent utilizing the shortest possible route. Except as provided in Section 504.3.3, the maximum vent connector horizontal length shall be 1 1/2 feet for each inch (457 mm per mm) of connector diameter as shown in Table 504.3.2.

TABLE 504.3.2
MAXIMUM VENT CONNECTOR LENGTH

CONNECTOR DIAMETER MAXIMUM (inches)	CONNECTOR HORIZONTAL LENGTH (feet)
3	4 1/2
4	6
5	7 1/2
6	9
7	10 1/2
8	12
9	13 1/2
10	15
12	18
14	21
16	24
18	27
20	30
22	33
24	36

For SI: 1 inch = 25.4 mm, 1 foot = 304.8 mm.

504.3.3 Connectors with longer lengths. Connectors with longer horizontal lengths than those listed in Section 504.3.2 are permitted under the following conditions:

1. The maximum capacity (FAN Max or NAT Max) of the vent connector shall be reduced 10 percent for each additional multiple of the length listed above. For example, the maximum length listed above for a 4-inch (102 mm) connector is 6 feet (1829 mm). With a connector length greater than 6 feet (1829 mm) but not exceeding 12 feet (3658 mm), the maximum capacity must be reduced by 10 percent (0.90 x maximum vent connector capacity). With a connector length greater than 12 feet (3658 mm) but not exceeding 18 feet (5486 mm), the maximum capacity must be reduced by 20 percent (0.80 x maximum vent capacity).

2. For a connector serving a fan-assisted appliance, the minimum capacity (FAN Min) of the connector shall be determined by referring to the corresponding single appliance table. For Type B double-wall connectors, Table 504.2(1) shall be used. For single-wall connectors, Table 504.2(2) shall be used. The height *(H)* and lateral *(L)* shall be measured according to the procedures for a single appliance vent, as if the other appliances were not present.

504.3.4 Combined connectors. Where the vent connectors are combined prior to entering the common vent, the maximum common vent capacity listed in the common venting tables shall be reduced 10 percent (0.90 (maximum common vent capacity). The length of the common vent connector manifold *(L_M)* shall not exceed 1 1/2 feet for each inch (457 mm per mm) of common vent connector manifold diameter *(D)*. (See Figure B-11.)

504.3.5 Common vertical vent offset. Where the common vertical vent is offset as shown in Figure B-12, the maximum common vent capacity listed in the common venting tables shall be reduced by 20 percent (0.80 x maximum common vent capacity), the equivalent of two 90-degree (1.6 rad) turns. The horizontal length of the common vent offset (L_M) shall not exceed $1^1/_2$ feet for each inch (457 mm per mm) of common vent diameter (D).

504.3.6 Additional capacity reduction. Excluding elbows counted in Section 504.3.5, for each additional 90-degree (1.6 rad) turn in excess of two, the maximum capacity of that portion of the venting system shall be reduced by 10 percent (0.90 x maximum common vent capacity). Two or more turns, the combined angles of which equal 90 degrees (1.6 rad), shall be considered equivalent to one 90-degree (1.6 rad) turn.

504.3.7 Common vent minimum size. The cross-sectional area of the common vent shall be equal to or greater than the cross-sectional area of the largest connector.

504.3.8 Common vent fittings. Interconnection fittings shall be the same size as the common vent.

504.3.9 High-altitude installations. Sea-level input ratings shall be used when determining maximum capacity for high-altitude installation. Actual input (derated for altitude) shall be used for determining minimum capacity for high-altitude installation.

504.3.10 Connector rise measurement. Connector rise (R) for each appliance connector shall be measured from the draft hood outlet or flue collar to the centerline where the vent gas streams come together.

504.3.11 Vent height measurement. For multiple units of equipment all located on one floor, available total height (H) shall be measured from the highest draft hood outlet or flue collar up to the level of the outlet of the common vent.

504.3.12 Multistory height measurement. For multistory installations, available total height (H) for each segment of the system shall be the vertical distance between the highest draft hood outlet or flue collar entering that segment and the centerline of the next higher interconnection tee (see Figure B-13).

504.3.13 Multistory lowest portion sizing. The size of the lowest connector and of the vertical vent leading to the lowest interconnection of a multistory system shall be in accordance with Table 504.2(1) or 504.2(2) for available total height (H) up to the lowest interconnection (see Figure B-14).

504.3.14 Multistory common vent offsets. Where used in multistory systems, vertical common vents shall be Type B double wall and shall be installed with a listed vent cap. A multistory common vertical vent shall be permitted to have a single offset, provided all of the following requirements are met:

1. The offset angle does not exceed 45 degrees (0.79 rad).
2. The horizontal length of the offset does not exceed $1^1/_2$ feet for each inch (457 mm per mm) of common vent diameter of the segment in which the offset is located.
3. For the segment of the common vertical vent containing the offset, the common vent capacity listed in the common venting tables is reduced by 20 percent (0.80 x maximum common vent capacity).
4. A multistory common vent shall not be reduced in size above the offset.

504.3.15 Vertical vent maximum size. Where two or more appliances are connected to a vertical vent or chimney, the flow area of the largest section of vertical vent or chimney shall not exceed seven times the smallest listed appliance categorized vent areas, flue collar area, or draft hood outlet area unless designed in accordance with approved engineering methods.

504.3.16 Multiple input rate appliances. For appliances with more than one input rate, the minimum vent connector capacity (FAN Min) determined from the tables shall be less than the lowest appliance input rating, and the maximum vent connector capacity (FAN Max or NAT Max) determined from the tables shall be greater than the highest appliance input rating.

504.3.17 Liner system sizing. Listed, corrugated metallic chimney liner systems in masonry chimneys shall be sized by using Table 504.3(1) or 504.3(2) for Type B vents, with the maximum capacity reduced by 20 percent (0.80 x maximum capacity) and the minimum capacity as shown in Table 504.3(1) or 504.3(2). Corrugated metallic liner systems installed with bends or offsets shall have their maximum capacity further reduced in accordance with Sections 504.3.5 and 504.3.6.

504.3.18 Chimney and vent location. Tables 504.3(1), 504.3(2), 504.3(3), 504.3(4), and 504.3(5) shall be used for chimneys and vents not exposed to the outdoors below the roof line. A Type B vent or listed chimney lining system passing through an unused masonry chimney flue shall not be considered to be exposed to the outdoors. Tables 504.3(7) and 504.3(8) shall be used for clay-tile-lined exterior masonry chimneys, provided all of the following conditions are met:

1. Vent connector is Type B double-wall.
2. At least one appliance is draft hood equipped.
3. The combined appliance input rating is less than the maximum capacity given by Table 504.3(7a) for NAT+NAT or Table 504.3(8a) for FAN+NAT.
4. The input rating of each space-heating appliance is

greater than the minimum input rating given by Table 504.3(7b) for NAT+NAT or Table 504.3(8b) for FAN+NAT.

5. The vent connector sizing is in accordance with Table 504.3(3).

Where these conditions cannot be met, an alternative venting design shall be used, such as a listed chimney lining system.

Exception: The installation of vents serving listed appliances shall be permitted to be in accordance with the appliance manufacturer's instructions and the terms of the listing.

504.3.19 Connector maximum size. Vent connectors shall not be increased in size more than two sizes greater than the listed appliance categorized vent diameter, flue collar diameter, or draft hood outlet diameter. Vent connectors for draft hood-equipped appliances shall not be smaller than the draft hood outlet diameter. Where a vent connector size(s) determined from the tables for a fan-assisted appliance(s) is smaller than the flue collar diameter, the smaller size(s) shall be permitted to be used provided all of the following conditions are met:

1. Vent connectors for fan-assisted appliance flue collars 12 inches (305 mm) in diameter or smaller are not reduced by more than one table size [e.g., 12 inches to 10 inches (305 mm to 254 mm) is a one-size reduction] and those larger than 12 inches (305 mm) in diameter are not reduced more than two table sizes [e.g., 24 inches to 20 inches (610 mm to 508 mm) is a two-size reduction].

2. The fan-assisted appliance(s) is common vented with a draft-hood-equipped appliances(s).

504.3.20 Component commingling. All combinations of pipe sizes, single-wall, and double-wall metal pipe shall be allowed within any connector run(s) or within the common vent, provided all of the appropriate tables permit all of the desired sizes and types of pipe, as if they were used for the entire length of the subject connector or vent. Where single-wall and Type B double-wall metal pipes are used for vent connectors, the common vent must be sized using Table 504.3(2) or 504.3(4), as appropriate.

504.3.21 Multiple sizes permitted. Where a table permits more than one diameter of pipe to be used for a connector or vent, all the permitted sizes shall be permitted to be used.

504.3.22 Table interpolation. Interpolation shall be permitted in calculating capacities for vent dimensions that fall between table entries (see Appendix B, Example 3).

504.3.23 Extrapolation prohibited. Extrapolation beyond the table entries shall not be permitted.

504.3.24 Engineering calculations. For vent heights less than 6 feet (1829 mm) and greater than shown in the tables, engineering methods shall be used to calculate vent capacities.

SECTION 505
DIRECT-VENT, INTEGRAL VENT, MECHANICAL VENT AND VENTILATION/EXHAUST HOOD VENTING

505.1 General. The installation of direct-vent and integral vent appliances shall be in accordance with Section 503. Mechanical venting systems and exhaust hood venting systems shall be designed and installed in accordance with Section 503.

SECTION 506
FACTORY-BUILT CHIMNEYS

506.1 Building heating appliances. Factory-built chimneys for building heating appliances producing flue gases having a temperature not greater than 1000°F (538°C), measured at the entrance to the chimney, shall be listed and labeled in accordance with UL 103 and shall be installed and terminated in accordance with the manufacturer's installation instructions.

506.2 Support. Where factory-built chimneys are supported by structural members, such as joists and rafters, such members shall be designed to support the additional load.

506.3 Medium-heat appliances. Factory-built chimneys for medium-heat appliances producing flue gases having a temperature above 1,000°F (538°C), measured at the entrance to the chimney, shall be listed and labeled in accordance with UL 959 and shall be installed and terminated in accordance with the manufacturer's installation instructions.

TABLE 504.2(1)
CAPACITY OF TYPE B DOUBLE-WALL GAS VENTS WHEN CONNECTED
DIRECTLY TO A SINGLE CATEGORY I APPLIANCE

		VENT DIAMETER (D)																				
		3"			4"			5"			6"			7"			8"			9"		
		APPLIANCE INPUT RATING IN THOUSANDS OF Btu/h																				
HEIGHT (H) (feet)	LATERAL (L) (feet)	FAN		NAT	FAN		NAT	FAN		NAT	FAN		NAT	FAN		NAT	FAN		NAT	FAN		NAT
		Min	Max	Max	Min	Max	Max	Min	Max	Max	Min	Max	Max	Min	Max	Max	Min	Max	Max	Min	Max	Max
6	0	0	78	46	0	152	86	0	251	141	0	375	205	0	524	285	0	698	370	0	897	470
	2	13	51	36	18	97	67	27	157	105	32	232	157	44	321	217	53	425	285	63	543	370
	4	21	49	34	30	94	64	39	153	103	50	227	153	66	316	211	79	419	279	93	536	362
	6	25	46	32	36	91	61	47	149	100	59	223	149	78	310	205	93	413	273	110	530	354
8	0	0	84	50	0	165	94	0	276	155	0	415	235	0	583	320	0	780	415	0	1,006	537
	2	12	57	40	16	109	75	25	178	120	28	263	180	42	365	247	50	483	322	60	619	418
	5	23	53	38	32	103	71	42	171	115	53	255	173	70	356	237	83	473	313	99	607	407
	8	28	49	35	39	98	66	51	164	109	64	247	165	84	347	227	99	463	303	117	596	396
10	0	0	88	53	0	175	100	0	295	166	0	447	255	0	631	345	0	847	450	0	1,096	585
	2	12	61	42	17	118	81	23	194	129	26	289	195	40	402	273	48	533	355	57	684	457
	5	23	57	40	32	113	77	41	187	124	52	280	188	68	392	263	81	522	346	95	671	446
	10	30	51	36	41	104	70	54	176	115	67	267	175	88	376	245	104	504	330	122	651	427
15	0	0	94	58	0	191	112	0	327	187	0	502	285	0	716	390	0	970	525	0	1,263	682
	2	11	69	48	15	136	93	20	226	150	22	339	225	38	475	316	45	633	414	53	815	544
	5	22	65	45	30	130	87	39	219	142	49	330	217	64	463	300	76	620	403	90	800	529
	10	29	59	41	40	121	82	51	206	135	64	315	208	84	445	288	99	600	386	116	777	507
	15	35	53	37	48	112	76	61	195	128	76	301	198	98	429	275	115	580	373	134	755	491
20	0	0	97	61	0	202	119	0	349	202	0	540	307	0	776	430	0	1,057	575	0	1,384	752
	2	10	75	51	14	149	100	18	250	166	20	377	249	33	531	346	41	711	470	50	917	612
	5	21	71	48	29	143	96	38	242	160	47	367	241	62	519	337	73	697	460	86	902	599
	10	28	64	44	38	133	89	50	229	150	62	351	228	81	499	321	95	675	443	112	877	576
	15	34	58	40	46	124	84	59	217	142	73	337	217	94	481	308	111	654	427	129	853	557
	20	48	52	35	55	116	78	69	206	134	84	322	206	107	464	295	125	634	410	145	830	537
30	0	0	100	64	0	213	128	0	374	220	0	587	336	0	853	475	0	1,173	650	0	1,548	855
	2	9	81	56	13	166	112	14	283	185	18	432	280	27	613	394	33	826	535	42	1,072	700
	5	21	77	54	28	160	108	36	275	176	45	421	273	58	600	385	69	811	524	82	1,055	688
	10	27	70	50	37	150	102	48	262	171	59	405	261	77	580	371	91	788	507	107	1,028	668
	15	33	64	NA	44	141	96	57	249	163	70	389	249	90	560	357	105	765	490	124	1,002	648
	20	56	58	NA	53	132	90	66	237	154	80	374	237	102	542	343	119	743	473	139	977	628
	30	NA	NA	NA	73	113	NA	88	214	NA	104	346	219	131	507	321	149	702	444	171	929	594
50	0	0	101	67	0	216	134	0	397	232	0	633	363	0	932	518	0	1,297	708	0	1,730	952
	2	8	86	61	11	183	122	14	320	206	15	497	314	22	715	445	26	975	615	33	1,276	813
	5	20	82	NA	27	177	119	35	312	200	43	487	308	55	702	438	65	960	605	77	1,259	798
	10	26	76	NA	35	168	114	45	299	190	56	471	298	73	681	426	86	935	589	101	1,230	773
	15	59	70	NA	42	158	NA	54	287	180	66	455	288	85	662	413	100	911	572	117	1,203	747
	20	NA	NA	NA	50	149	NA	63	275	169	76	440	278	97	642	401	113	888	556	131	1,176	722
	30	NA	NA	NA	69	131	NA	84	250	NA	99	410	259	123	605	376	141	844	522	161	1,125	670
100	0	NA	NA	NA	0	218	NA	0	407	NA	0	665	400	0	997	560	0	1,411	770	0	1,908	1,040
	2	NA	NA	NA	10	194	NA	12	354	NA	13	566	375	18	831	510	21	1,155	700	25	1,536	935
	5	NA	NA	NA	26	189	NA	33	347	NA	40	557	369	52	820	504	60	1,141	692	71	1,519	926
	10	NA	NA	NA	33	182	NA	43	335	NA	53	542	361	68	801	493	80	1,118	679	94	1,492	910
	15	NA	NA	NA	40	174	NA	50	321	NA	62	528	353	80	782	482	93	1,095	666	109	1,465	895
	20	NA	NA	NA	47	166	NA	59	311	NA	71	513	344	90	763	471	105	1,073	653	122	1,438	880
	30	NA	NA	NA	NA	NA	NA	78	290	NA	92	483	NA	115	726	449	131	1,029	627	149	1,387	849
	50	NA	NA	NA	NA	NA	NA	NA	NA	NA	147	428	NA	180	651	405	197	944	575	217	1,288	787

(continued)

TABLE 504.2(1)   CHIMNEYS AND VENTS

TABLE 504.2(1)—continued
CAPACITY OF TYPE B DOUBLE-WALL GAS VENTS WHEN CONNECTED DIRECTLY TO A SINGLE CATEGORY I APPLIANCE

		VENT DIAMETER (D)																							
		10"			12"			14"			16"			18"			20"			22"			24"		
		APPLIANCE INPUT RATING IN THOUSANDS OF Btu/h																							
HEIGHT (H)	LATERAL (L)	FAN		NAT	FAN		NAT	FAN		NAT	FAN		NAT	FAN		NAT	FAN		NAT	FAN		NAT	FAN		NAT
(feet)	(feet)	Min	Max	Max	Min	Max	Max	Min	Max	Max	Min	Max	Max	Min	Max	Max	Min	Max	Max	Min	Max	Max	Min	Max	Max
6	0	0	1,121	570	0	1,645	850	0	2,267	1,170	0	2,983	1,530	0	3,802	1,960	0	4,721	2,430	0	5,737	2,950	0	6,853	3,520
	2	75	675	455	103	982	650	138	1,346	890	178	1,769	1,170	225	2,250	1,480	296	2,782	1,850	360	3,377	2,220	426	4,030	2,670
	4	110	668	445	147	975	640	191	1,338	880	242	1,761	1,160	300	2,242	1,475	390	2,774	1,835	469	3,370	2,215	555	4,023	2,660
	6	128	661	435	171	967	630	219	1,330	870	276	1,753	1,150	341	2,235	1,470	437	2,767	1,820	523	3,363	2,210	618	4,017	2,650
8	0	0	1,261	660	0	1,858	970	0	2,571	1,320	0	3,399	1,740	0	4,333	2,220	0	5,387	2,750	0	6,555	3,360	0	7,838	4,010
	2	71	770	515	98	1,124	745	130	1,543	1,020	168	2,030	1,340	212	2,584	1,700	278	3,196	2,110	336	3,882	2,560	401	4,634	3,050
	5	115	758	503	154	1,110	733	199	1,528	1,010	251	2,013	1,330	311	2,563	1,685	398	3,180	2,090	476	3,863	2,545	562	4,612	3,040
	8	137	746	490	180	1,097	720	231	1,514	1,000	289	2,000	1,320	354	2,552	1,670	450	3,163	2,070	537	3,850	2,530	630	4,602	3,030
10	0	0	1,377	720	0	2,036	1,060	0	2,825	1,450	0	3,742	1,925	0	4,782	2,450	0	5,955	3,050	0	7,254	3,710	0	8,682	4,450
	2	68	852	560	93	1,244	850	124	1,713	1,130	161	2,256	1,480	202	2,868	1,890	264	3,556	2,340	319	4,322	2,840	378	5,153	3,390
	5	112	839	547	149	1,229	829	192	1,696	1,105	243	2,238	1,461	300	2,849	1,871	382	3,536	2,318	458	4,301	2,818	540	5,132	3,371
	10	142	817	525	187	1,204	795	238	1,669	1,080	298	2,209	1,430	364	2,818	1,840	459	3,504	2,280	546	4,268	2,780	641	5,099	3,340
15	0	0	1,596	840	0	2,380	1,240	0	3,323	1,720	0	4,423	2,270	0	5,678	2,900	0	7,099	3,620	0	8,665	4,410	0	10,393	5,300
	2	63	1,019	675	86	1,495	985	114	2,062	1,350	147	2,719	1,770	186	3,467	2,260	239	4,304	2,800	290	5,232	3,410	346	6,251	4,080
	5	105	1,003	660	140	1,476	967	182	2,041	1,327	229	2,696	1,748	283	3,442	2,235	355	4,278	2,777	426	5,204	3,385	501	6,222	4,057
	10	135	977	635	177	1,446	936	227	2,009	1,289	283	2,659	1,712	346	3,402	2,193	432	4,234	2,739	510	5,159	3,343	599	6,175	4,019
	15	155	953	610	202	1,418	905	257	1,976	1,250	318	2,623	1,675	385	3,363	2,150	479	4,192	2,700	564	5,115	3,300	665	6,129	3,980
20	0	0	1,756	930	0	2,637	1,350	0	3,701	1,900	0	4,948	2520	0	6,376	3,250	0	7,988	4,060	0	9,785	4,980	0	11,753	6,000
	2	59	1,150	755	81	1,694	1,100	107	2,343	1,520	139	3,097	2,000	175	3,955	2,570	220	4,916	3,200	269	5,983	3,910	321	7,154	4,700
	5	101	1,133	738	135	1,674	1,079	174	2,320	1,498	219	3,071	1,978	270	3,926	2,544	337	4,885	3,174	403	5,950	3,880	475	7,119	4,662
	10	130	1,105	710	172	1,641	1,045	220	2,282	1,460	273	3,029	1,940	334	3,880	2,500	413	4,835	3,130	489	5,896	3,830	573	7,063	4,600
	15	150	1,078	688	195	1,609	1,018	248	2,245	1,425	306	2,988	1,910	372	3,835	2,465	459	4,786	3,090	541	5,844	3,795	631	7,007	4,575
	20	167	1,052	665	217	1,578	990	273	2,210	1,390	335	2,948	1,880	404	3,791	2,430	495	4,737	3,050	585	5,792	3,760	689	6,953	4,550
30	0	0	1,977	1,060	0	3,004	1,550	0	4,252	2,170	0	5,725	2,920	0	7,420	3,770	0	9,341	4,750	0	11,483	5,850	0	13,848	7,060
	2	54	1,351	865	74	2,004	1,310	98	2,786	1,800	127	3,696	2,380	159	4,734	3,050	199	5,900	3,810	241	7,194	4,650	285	8,617	5,600
	5	96	1,332	851	127	1,981	1,289	164	2,759	1,775	206	3,666	2,350	252	4,701	3,020	312	5,863	3,783	373	7,155	4,622	439	8,574	5,552
	10	125	1,301	829	164	1,944	1,254	209	2,716	1,733	259	3,617	2,300	316	4,647	2,970	386	5,803	3,739	456	7,090	4,574	535	8,505	5,471
	15	143	1,272	807	187	1,908	1,220	237	2,674	1,692	292	3,570	2,250	354	4,594	2,920	431	5,744	3,695	507	7,026	4,527	590	8,437	5,391
	20	160	1,243	784	207	1,873	1,185	260	2,633	1,650	319	3,523	2,200	384	4,542	2,870	467	5,686	3,650	548	6,964	4,480	639	8,370	5,310
	30	195	1,189	745	246	1,807	1,130	305	2,555	1,585	369	3,433	2,130	440	4,442	2,785	540	5,574	3,565	635	6,842	4,375	739	8,239	5,225
50	0	0	2,231	1,195	0	3,441	1,825	0	4,934	2,550	0	6,711	3,440	0	8,774	4,460	0	11,129	5,635	0	13,767	6,940	0	16,694	8,430
	2	41	1,620	1,010	66	2,431	1,513	86	3,409	2,125	113	4,554	2,840	141	5,864	3,670	171	7,339	4,630	209	8,980	5,695	251	10,788	6,860
	5	90	1,600	996	118	2,406	1,495	151	3,380	2,102	191	4,520	2,813	234	5,826	3,639	283	7,295	4,597	336	8,933	5,654	394	10,737	6,818
	10	118	1,567	972	154	2,366	1,466	196	3,332	2,064	243	4,464	2,767	295	5,763	3,585	355	7,224	4,542	419	8,855	5,585	491	10,652	6,749
	15	136	1,536	948	177	2,327	1,437	222	3,285	2,026	274	4,409	2,721	330	5,701	3,534	396	7,155	4,511	465	8,779	5,546	542	10,570	6,710
	20	151	1,505	924	195	2,288	1,408	244	3,239	1,987	300	4,356	2,675	361	5,641	3,481	433	7,086	4,479	506	8,704	5,506	586	10,488	6,670
	30	183	1,446	876	232	2,214	1,349	287	3,150	1,910	347	4,253	2,631	412	5,523	3,431	494	6,953	4,421	577	8,557	5,444	672	10,328	6,603
100	0	0	2,491	1,310	0	3,925	2,050	0	5,729	2,950	0	7,914	4,050	0	10,485	5,300	0	13,454	6,700	0	16,817	8,600	0	20,578	10,300
	2	30	1,975	1,170	44	3,027	1,820	72	4,313	2,550	95	5,834	3,500	120	7,591	4,600	138	9,577	5,800	169	11,803	7,200	204	14,264	8,800
	5	82	1,955	1,159	107	3,002	1,803	136	4,282	2,531	172	5,797	3,475	208	7,548	4,566	245	9,528	5,769	293	11,748	7,162	341	14,204	8,756
	10	108	1,923	1,142	142	2,961	1,775	180	4,231	2,500	223	5,737	3,434	268	7,478	4,509	318	9,447	5,717	374	11,658	7,100	436	14,105	8,683
	15	126	1,892	1,124	163	2,920	1,747	206	4,182	2,469	252	5,678	3,392	304	7,409	4,451	358	9,367	5,665	418	11,569	7,037	487	14,007	8,610
	20	141	1,861	1,107	181	2,880	1,719	226	4,133	2,438	277	5,619	3,351	330	7,341	4,394	387	9,289	5,613	452	11,482	6,975	523	13,910	8,537
	30	170	1,802	1,071	215	2,803	1,663	265	4,037	2,375	319	5,505	3,267	378	7,209	4,279	446	9,136	5,509	514	11,310	6,850	592	13,720	8,391
	50	241	1,688	1,000	292	2,657	1,550	350	3,856	2,250	415	5,289	3,100	486	6,956	4,050	572	8,841	5,300	659	10,979	6,600	752	13,354	8,100

For SI: 1 inch = 25.4 mm, 1 foot = 304.8 mm, 1 British thermal unit per hour = 0.2931 W.

TABLE 504.2(2)

TABLE 504.2(2)
CAPACITY OF TYPE B DOUBLE-WALL VENTS WITH SINGLE-WALL METAL CONNECTORS SERVING A SINGLE CATEGORY I APPLIANCE

VENT DIAMETER (D) — APPLIANCE INPUT RATING IN THOUSANDS OF Btu/h

H (feet)	L (feet)	3" FAN Min	3" FAN Max	3" NAT Max	4" FAN Min	4" FAN Max	4" NAT Max	5" FAN Min	5" FAN Max	5" NAT Max	6" FAN Min	6" FAN Max	6" NAT Max	7" FAN Min	7" FAN Max	7" NAT Max	8" FAN Min	8" FAN Max	8" NAT Max	9" FAN Min	9" FAN Max	9" NAT Max	10" FAN Min	10" FAN Max	10" NAT Max	12" FAN Min	12" FAN Max	12" NAT Max
6	0	38	77	45	59	151	85	85	249	140	126	373	204	165	522	284	211	695	369	267	894	469	371	1,118	569	537	1,639	849
	2	39	51	36	60	96	66	85	156	104	123	231	156	159	320	213	201	423	284	251	541	368	347	673	453	498	979	648
	4	NA	NA	33	74	92	63	102	152	102	146	225	152	187	313	208	237	416	277	295	533	360	409	664	443	584	971	638
	6	NA	NA	31	83	89	60	114	147	99	163	220	148	207	307	203	263	409	271	327	526	352	449	656	433	638	962	627
8	0	37	83	50	58	164	93	83	273	154	123	412	234	161	580	319	206	777	414	258	1,002	536	360	1,257	658	521	1,852	967
	2	39	56	39	59	108	75	83	176	119	121	261	179	155	363	246	197	482	321	246	617	417	339	768	513	486	1,120	743
	5	NA	NA	37	77	102	69	107	168	114	151	252	171	193	352	235	245	470	311	305	604	404	418	754	500	598	1,104	730
	8	NA	NA	33	90	95	64	122	161	107	175	243	163	223	342	225	280	458	300	344	591	392	470	740	486	665	1,089	715
10	0	37	87	53	57	174	99	82	293	165	120	444	254	158	628	344	202	844	449	253	1,093	584	351	1,373	718	507	2,031	1,057
	2	39	61	41	59	117	80	82	193	128	119	287	194	153	400	272	193	531	354	242	681	456	332	849	559	475	1,242	848
	5	52	56	39	76	111	76	105	185	122	148	277	186	190	388	261	241	518	344	299	667	443	409	834	544	584	1,224	825
	10	NA	NA	34	97	100	68	132	171	112	188	261	171	237	369	241	296	497	325	363	643	423	492	808	520	688	1,194	788
15	0	36	93	57	56	190	111	80	325	186	116	499	283	153	713	388	195	966	523	244	1,259	681	336	1,591	838	488	2,374	1,237
	2	38	69	47	57	136	93	80	225	149	115	337	224	148	473	314	187	631	413	232	812	543	319	1,015	673	457	1,491	983
	5	51	63	44	75	128	86	102	216	140	144	326	217	182	459	298	231	616	400	287	795	526	392	997	657	562	1,469	963
	10	NA	NA	39	95	116	79	128	201	131	182	308	203	228	438	284	284	592	381	349	768	501	470	966	628	664	1,433	928
	15	NA	NA	NA	NA	NA	72	158	186	124	220	290	192	272	418	269	334	568	367	404	742	484	540	937	601	750	1,399	894
20	0	35	96	60	54	200	118	78	346	201	114	537	306	149	772	428	190	1,053	573	238	1,379	750	326	1,751	927	473	2,631	1,346
	2	37	74	50	56	148	99	78	248	165	113	375	248	144	528	344	182	708	468	227	914	611	309	1,146	754	443	1,689	1,098
	5	50	68	47	73	140	94	100	239	158	141	363	239	178	514	334	224	692	457	279	896	596	381	1,126	734	547	1,665	1,074
	10	NA	NA	41	93	129	86	125	223	146	177	344	224	222	491	316	277	666	437	339	866	570	457	1,092	702	646	1,626	1,037
	15	NA	NA	NA	NA	NA	80	155	208	136	216	325	210	264	469	301	325	640	419	393	838	549	526	1,060	677	730	1,587	1,005
	20	NA	NA	NA	NA	NA	NA	186	192	126	254	306	196	309	448	285	374	616	400	448	810	526	592	1,028	651	808	1,550	973
30	0	34	99	63	53	211	127	76	372	219	110	584	334	144	849	472	184	1,168	647	229	1,542	852	312	1,971	1,056	454	2,996	1,545
	2	37	80	56	55	164	111	76	281	183	109	429	279	139	610	392	175	823	533	219	1,069	698	296	1,346	863	424	1,999	1,308
	5	49	74	52	72	157	106	98	271	173	136	417	271	171	595	382	215	806	521	269	1,049	684	366	1,324	846	524	1,971	1,283
	10	NA	NA	NA	91	144	98	122	255	168	171	397	257	213	570	367	265	777	501	327	1,017	662	440	1,287	821	620	1,927	1,243
	15	NA	NA	NA	115	131	NA	151	239	157	208	377	242	255	547	349	312	750	481	379	985	638	507	1,251	794	702	1,884	1,205
	20	NA	NA	NA	NA	NA	NA	181	223	NA	246	357	228	298	524	333	360	723	461	433	955	615	570	1,216	768	780	1,841	1,166
	30	NA	NA	NA	NA	NA	NA	NA	NA	NA	NA	NA	NA	389	477	305	461	670	426	541	895	574	704	1,147	720	937	1,759	1,101
50	0	33	99	66	51	213	133	73	394	230	105	629	361	138	928	515	176	1,292	704	220	1,724	948	295	2,223	1,189	428	3,432	1,818
	2	36	84	61	53	181	121	73	318	205	104	495	312	133	712	443	168	971	613	209	1,273	811	280	1,615	1,007	401	2,426	1,509
	5	48	80	NA	70	174	117	94	308	198	131	482	305	164	696	435	204	953	602	257	1,252	795	347	1,591	991	496	2,396	1,490
	10	NA	NA	NA	89	160	NA	118	292	186	162	461	292	203	671	420	253	923	583	313	1,217	765	418	1,551	963	589	2,347	1,455
	15	NA	NA	NA	112	148	NA	145	275	174	199	441	280	244	646	405	299	894	562	363	1,183	736	481	1,512	934	668	2,299	1,421
	20	NA	NA	NA	NA	NA	NA	176	257	NA	236	420	267	285	622	389	345	866	543	415	1,150	708	544	1,473	906	741	2,251	1,387
	30	NA	NA	NA	NA	NA	NA	NA	NA	NA	315	376	NA	373	573	NA	442	809	502	521	1,086	649	674	1,399	848	892	2,159	1,318
100	0	NA	NA	NA	49	214	NA	69	403	NA	100	659	395	131	991	555	166	1,404	765	207	1,900	1,033	273	2,479	1,300	395	3,912	2,042
	2	NA	NA	NA	51	192	NA	70	351	NA	98	563	373	125	828	508	158	1,152	698	196	1,532	933	259	1,970	1,168	371	3,021	1,817
	5	NA	NA	NA	67	186	NA	90	342	NA	125	551	366	156	813	501	194	1,134	688	240	1,511	921	322	1,945	1,153	460	2,990	1,796
	10	NA	NA	NA	85	175	NA	113	324	NA	153	532	354	191	789	486	238	1,104	672	293	1,477	902	389	1,905	1,133	547	2,938	1,763
	15	NA	NA	NA	132	162	NA	138	310	NA	188	511	343	230	764	473	281	1,075	656	342	1,443	884	447	1,865	1,110	618	2,888	1,730
	20	NA	NA	NA	NA	NA	NA	168	295	NA	224	487	NA	270	739	458	325	1,046	639	391	1,410	864	507	1,825	1,087	690	2,838	1,696
	30	NA	NA	NA	NA	NA	NA	231	264	NA	301	448	NA	355	685	NA	418	988	NA	491	1,343	824	631	1,747	1,041	834	2,739	1,627
	50	NA	NA	NA	NA	A	NA	NA	NA	NA	NA	NA	NA	540	584	NA	617	866	NA	711	1,205	NA	895	1,591	NA	1,138	2,547	1,489

For SI: 1 inch = 25.4 mm, 1 foot = 304.8 mm, 1 British thermal unit per hour = 0.2931 W.

TABLE 504.2(3)   CHIMNEYS AND VENTS

TABLE 504.2(3)
CAPACITY OF MASONRY CHIMNEY FLUE WITH TYPE B DOUBLE-WALL VENT CONNECTORS SERVING A SINGLE CATEGORY I APPLIANCE

TYPE B DOUBLE-WALL CONNECTOR DIAMETER (D)
To be used with chimney areas within the size limits at bottom

APPLIANCE INPUT RATING IN THOUSANDS OF Btu/h

		3"			4"			5"			6"			7"			8"			9"			10"			12"		
		FAN		NAT	FAN		NAT	FAN		NAT	FAN		NAT	FAN		NAT	FAN		NAT	FAN		NAT	FAN		NAT	FAN		NAT
HEIGHT (H) (feet)	LATERAL (L) (feet)	Min	Max	Max	Min	Max	Max	Min	Max	Max	Min	Max	Max	Min	Max	Max	Min	Max	Max	Min	Max	Max	Min	Max	Max	Min	Max	Max
6	2	NA	NA	28	NA	NA	52	NA	NA	86	NA	NA	130	NA	NA	180	NA	NA	247	NA	NA	320	NA	NA	401	NA	NA	581
	5	NA	NA	25	NA	NA	49	NA	NA	82	NA	NA	117	NA	NA	165	NA	NA	231	NA	NA	298	NA	NA	376	NA	NA	561
8	2	NA	NA	29	NA	NA	55	NA	NA	93	NA	NA	145	NA	NA	198	NA	NA	266	84	590	350	100	728	446	139	1,024	651
	5	NA	NA	26	NA	NA	52	NA	NA	88	NA	NA	134	NA	NA	183	NA	NA	247	NA	NA	328	149	711	423	201	1,007	640
	8	NA	NA	24	NA	NA	48	NA	NA	83	NA	NA	127	NA	NA	175	NA	NA	239	NA	NA	318	173	695	410	231	990	623
10	2	NA	NA	31	NA	NA	61	NA	NA	103	NA	NA	162	NA	NA	221	68	519	298	82	655	388	98	810	491	136	1,144	724
	5	NA	NA	28	NA	NA	57	NA	NA	96	NA	NA	148	NA	NA	204	NA	NA	277	124	638	365	146	791	466	196	1,124	712
	10	NA	NA	25	NA	NA	50	NA	NA	87	NA	NA	139	NA	NA	191	NA	NA	263	155	610	347	182	762	444	240	1,093	668
15	2	NA	NA	35	NA	NA	67	NA	NA	114	NA	NA	179	53	475	250	64	613	336	77	779	441	92	968	562	127	1,376	841
	5	NA	NA	35	NA	NA	62	NA	NA	107	NA	NA	164	NA	NA	231	99	594	313	118	759	416	139	946	533	186	1,352	828
	10	NA	NA	28	NA	NA	55	NA	NA	97	NA	NA	153	NA	NA	216	126	565	296	148	727	394	173	912	567	229	1,315	777
	15	NA	NA	NA	NA	NA	48	NA	NA	89	NA	NA	141	NA	NA	201	NA	NA	281	171	698	375	198	880	485	259	1,280	742
20	2	NA	NA	38	NA	NA	74	NA	NA	124	NA	NA	201	51	522	274	61	678	375	73	867	491	87	1,083	627	121	1,548	953
	5	NA	NA	36	NA	NA	68	NA	NA	116	NA	NA	184	80	503	254	95	658	350	113	845	463	133	1,059	597	179	1,523	933
	10	NA	NA	NA	NA	NA	60	NA	NA	107	NA	NA	172	NA	NA	237	122	627	332	143	811	440	167	1,022	566	221	1,482	879
	15	NA	NA	NA	NA	NA	NA	NA	NA	97	NA	NA	159	NA	NA	220	NA	NA	314	165	780	418	191	987	541	251	1,443	840
	20	NA	NA	NA	NA	NA	NA	NA	NA	83	NA	NA	148	NA	NA	206	NA	NA	296	186	750	397	214	955	513	277	1,406	807
30	2	NA	NA	41	NA	NA	82	NA	NA	137	NA	NA	216	47	581	303	57	762	421	68	985	558	81	1,240	717	111	1,793	1,112
	5	NA	NA	NA	NA	NA	76	NA	NA	128	NA	NA	198	75	561	281	90	741	393	106	962	526	125	1,216	683	169	1,766	1,094
	10	NA	NA	NA	NA	NA	67	NA	NA	115	NA	NA	184	NA	NA	263	115	709	373	135	927	500	158	1,176	648	210	1,721	1,025
	15	NA	NA	NA	NA	NA	NA	NA	NA	107	NA	NA	171	NA	NA	243	NA	NA	353	156	893	476	181	1,139	621	239	1,679	981
	20	NA	NA	NA	NA	NA	NA	NA	NA	91	NA	NA	159	NA	NA	227	NA	NA	332	176	860	450	203	1,103	592	264	1,638	940
	30	NA	NA	NA	NA	NA	NA	NA	NA	NA	NA	NA	NA	NA	NA	188	NA	NA	288	NA	NA	416	249	1,035	555	318	1,560	877
50	2	NA	NA	NA	NA	NA	92	NA	NA	161	NA	NA	251	NA	NA	351	51	840	477	61	1,106	633	72	1,413	812	99	2,080	1,243
	5	NA	NA	NA	NA	NA	NA	NA	NA	151	NA	NA	230	NA	NA	323	83	819	445	98	1,083	596	116	1,387	774	155	2,052	1,225
	10	NA	NA	NA	NA	NA	NA	NA	NA	138	NA	NA	215	NA	NA	304	NA	NA	424	126	1,047	567	147	1,347	733	195	2,006	1,147
	15	NA	NA	NA	NA	NA	NA	NA	NA	127	NA	NA	199	NA	NA	282	NA	NA	400	146	1,010	539	170	1,307	702	222	1,961	1,099
	20	NA	NA	NA	NA	NA	NA	NA	NA	NA	NA	NA	185	NA	NA	264	NA	NA	376	165	977	511	190	1,269	669	246	1,916	1,050
	30	NA	NA	NA	NA	NA	NA	NA	NA	NA	NA	NA	NA	NA	NA	NA	NA	NA	327	NA	NA	468	233	1,196	623	295	1,832	984
Minimum Internal Area of Chimney square inches		12			19			28			38			50			63			78			95			132		
Maximum Internal Area of Chimney square inches		49			88			137			198			269			352			445			550			792		

For SI: 1 inch = 25.4 mm, 1 square inch = 645.16 mm², 1 foot = 304.8 mm, 1 British thermal unit per hour = 0.2931 W.

TABLE 504.2(4)
CAPACITY OF MASONRY CHIMNEY FLUE WITH SINGLE-WALL VENT CONNECTORS SERVING A SINGLE CATEGORY I APPLIANCE

SINGLE-WALL METAL CONNECTOR DIAMETER (D)
To be used with chimney areas within the size limits at bottom

APPLIANCE INPUT RATING IN THOUSANDS OF Btu/h

HEIGHT (H) (feet)	LATERAL (L) (feet)	3" FAN Min	3" FAN Max	3" NAT Max	4" FAN Min	4" FAN Max	4" NAT Max	5" FAN Min	5" FAN Max	5" NAT Max	6" FAN Min	6" FAN Max	6" NAT Max	7" FAN Min	7" FAN Max	7" NAT Max	8" FAN Min	8" FAN Max	8" NAT Max	9" FAN Min	9" FAN Max	9" NAT Max	10" FAN Min	10" FAN Max	10" NAT Max	12" FAN Min	12" FAN Max	12" NAT Max
6	2	NA	NA	28	NA	NA	52	NA	NA	86	NA	NA	130	NA	NA	180	NA	NA	247	NA	NA	319	NA	NA	400	NA	NA	580
	5	NA	NA	25	NA	NA	48	NA	NA	81	NA	NA	116	NA	NA	164	NA	NA	230	NA	NA	297	NA	NA	375	NA	NA	560
8	2	NA	NA	29	NA	NA	55	NA	NA	93	NA	NA	145	NA	NA	197	NA	NA	265	NA	NA	349	382	725	445	549	1,021	650
	5	NA	NA	26	NA	NA	51	NA	NA	87	NA	NA	133	NA	NA	182	NA	NA	246	NA	NA	327	NA	NA	422	673	1,003	638
	8	NA	NA	23	NA	NA	47	NA	NA	82	NA	NA	126	NA	NA	174	NA	NA	237	NA	NA	317	NA	NA	408	747	985	621
10	2	NA	NA	31	NA	NA	61	NA	NA	102	NA	NA	161	NA	NA	220	216	518	297	271	654	387	373	808	490	536	1,142	722
	5	NA	NA	28	NA	NA	56	NA	NA	95	NA	NA	147	NA	NA	203	NA	NA	276	334	635	364	459	789	465	657	1,121	710
	10	NA	NA	24	NA	NA	49	NA	NA	86	NA	NA	137	NA	NA	189	NA	NA	261	NA	NA	345	547	758	441	771	1,088	665
15	2	NA	NA	35	NA	NA	67	NA	NA	113	NA	NA	178	166	473	249	211	611	335	264	776	440	362	965	560	520	1,373	840
	5	NA	NA	32	NA	NA	61	NA	NA	106	NA	NA	163	NA	NA	230	261	591	312	325	755	414	444	942	531	637	1,348	825
	10	NA	NA	27	NA	NA	54	NA	NA	96	NA	NA	151	NA	NA	214	NA	NA	294	392	722	392	531	907	504	749	1,309	774
	15	NA	NA	NA	NA	NA	46	NA	NA	87	NA	NA	138	NA	NA	198	NA	NA	278	452	692	372	606	873	481	841	1,272	738
20	2	NA	NA	38	NA	NA	73	NA	NA	123	NA	NA	200	163	520	273	206	675	374	258	864	490	252	1,079	625	508	1,544	950
	5	NA	NA	35	NA	NA	67	NA	NA	115	NA	NA	183	NA	NA	252	255	655	348	317	842	461	433	1,055	594	623	1,518	930
	10	NA	NA	NA	NA	NA	59	NA	NA	105	NA	NA	170	NA	NA	235	312	622	330	382	806	437	517	1,016	562	733	1,475	875
	15	NA	NA	NA	NA	NA	NA	NA	NA	95	NA	NA	156	NA	NA	217	NA	NA	311	442	773	414	591	979	539	823	1,434	835
	20	NA	NA	NA	NA	NA	NA	NA	NA	80	NA	NA	144	NA	NA	202	NA	NA	292	NA	NA	392	663	944	510	911	1,394	800
30	2	NA	NA	41	NA	NA	81	NA	NA	136	NA	NA	215	158	578	302	200	759	420	249	982	556	340	1,237	715	489	1,789	1,110
	5	NA	NA	NA	NA	NA	75	NA	NA	127	NA	NA	196	NA	NA	279	245	737	391	306	958	524	417	1,210	680	600	1,760	1,090
	10	NA	NA	NA	NA	NA	66	NA	NA	113	NA	NA	182	NA	NA	260	300	703	370	370	920	496	500	1,168	644	708	1,713	1,020
	15	NA	NA	NA	NA	NA	NA	NA	NA	105	NA	NA	168	NA	NA	240	NA	NA	349	428	884	471	572	1,128	615	798	1,668	975
	20	NA	NA	NA	NA	NA	NA	NA	NA	88	NA	NA	155	NA	NA	223	NA	NA	327	NA	NA	445	643	1,089	585	883	1,624	932
	30	NA	NA	NA	NA	NA	NA	NA	NA	NA	NA	NA	NA	NA	NA	182	NA	NA	281	NA	NA	408	NA	NA	544	1,055	1,539	865
50	2	NA	NA	NA	NA	NA	91	NA	NA	160	NA	NA	250	NA	NA	350	191	837	475	238	1,103	631	323	1,408	810	463	2,076	1,240
	5	NA	NA	NA	NA	NA	NA	NA	NA	149	NA	NA	228	NA	NA	321	NA	NA	442	293	1,078	593	398	1,381	770	571	2,044	1,220
	10	NA	NA	NA	NA	NA	NA	NA	NA	136	NA	NA	212	NA	NA	301	NA	NA	420	355	1,038	562	447	1,337	728	674	1,994	1,140
	15	NA	NA	NA	NA	NA	NA	NA	NA	124	NA	NA	195	NA	NA	278	NA	NA	395	NA	NA	533	546	1,294	695	761	1,945	1,090
	20	NA	NA	NA	NA	NA	NA	NA	NA	NA	NA	NA	180	NA	NA	258	NA	NA	370	NA	NA	504	616	1,251	660	844	1,898	1,040
	30	NA	NA	NA	NA	NA	NA	NA	NA	NA	NA	NA	NA	NA	NA	NA	NA	NA	318	NA	NA	458	NA	NA	610	1,009	1,805	970
Minimum Internal Area of Chimney Square Inches		12			19			28			38			50			63			78			95			132		
Maximum Internal Area of Chimney Square Inches		49			88			137			198			269			352			445			550			792		

For SI: 1 inch = 25.4 mm, 1 square inch = 645.16 mm^2, 1 foot = 304.8 mm, 1 British thermal unit per hour = 0.2931 W.

TABLE 504.2(5)

CHIMNEYS AND VENTS

TABLE 504.2(5)
CAPACITY OF SINGLE-WALL METAL PIPE OR TYPE B ASBESTOS CEMENT VENTS
SERVING A SINGLE DRAFT HOOD EQUIPPED APPLIANCE

HEIGHT (H) (feet)	LATERAL (L) (feet)	VENT DIAMETER (D)							
		3"	4"	5"	6"	7"	8"	10"	12"
		MAXIMUM APPLIANCE INPUT RATING IN THOUSANDS OF Btu/h							
6	0	39	70	116	170	232	312	500	750
	2	31	55	94	141	194	260	415	620
	5	28	51	88	128	177	242	390	600
8	0	42	76	126	185	252	340	542	815
	2	32	61	102	154	210	284	451	680
	5	29	56	95	141	194	264	430	648
	10	24	49	86	131	180	250	406	625
10	0	45	84	138	202	279	372	606	912
	2	35	67	111	168	233	311	505	760
	5	32	61	104	153	215	289	480	724
	10	27	54	94	143	200	274	455	700
	15	NA	46	84	130	186	258	432	666
15	0	49	91	151	223	312	420	684	1040
	2	39	72	122	186	260	350	570	865
	5	35	67	110	170	240	325	540	825
	10	30	58	103	158	223	308	514	795
	15	NA	50	93	144	207	291	488	760
	20	NA	NA	82	132	195	273	466	726
20	0	53	101	163	252	342	470	770	1190
	2	42	80	136	210	286	392	641	990
	5	38	74	123	192	264	364	610	945
	10	32	65	115	178	246	345	571	910
	15	NA	55	104	163	228	326	550	870
	20	NA	NA	91	149	214	306	525	832
30	0	56	108	183	276	384	529	878	1370
	2	44	84	148	230	320	441	730	1140
	5	NA	78	137	210	296	410	694	1080
	10	NA	68	125	196	274	388	656	1050
	15	NA	NA	113	177	258	366	625	1000
	20	NA	NA	99	163	240	344	596	960
	30	NA	NA	NA	NA	192	295	540	890
50	0	NA	120	210	310	443	590	980	1550
	2	NA	95	171	260	370	492	820	1290
	5	NA	NA	159	234	342	474	780	1230
	10	NA	NA	146	221	318	456	730	1190
	15	NA	NA	NA	200	292	407	705	1130
	20	NA	NA	NA	185	276	384	670	1080
	30	NA	NA	NA	NA	222	330	605	1010

For SI: 1 inch = 25.4 mm, 1 foot = 304.8 mm, 1 British thermal unit per hour = 0.2931 W.

TABLE 504.3(1)
CAPACITY OF TYPE B DOUBLE-WALL VENTS WITH TYPE B DOUBLE-WALL CONNECTORS SERVING TWO OR MORE CATEGORY I APPLIANCES

VENT CONNECTOR CAPACITY

		TYPE B DOUBLE-WALL VENT AND CONNECTOR DIAMETER (D)																						
		3"			4"			5"			6"			7"			8"			9"			10"		
		APPLIANCE INPUT RATING LIMITS IN THOUSANDS OF Btu/h																						
Vent Height	Connector Rise	FAN		NAT	FAN		NAT	FAN		NAT	FAN		NAT	FAN		NAT	FAN		NAT	FAN		NAT	FAN		NAT
(H) (feet)	(R) (feet)	Min	Max	Max	Min	Max	Max	Min	Max	Max	Min	Max	Max	Min	Max	Max	Min	Max	Max	Min	Max	Max	Min	Max	Max
6	1	22	37	26	35	66	46	46	106	72	58	164	104	77	225	142	92	296	185	109	376	237	128	466	289
	2	23	41	31	37	75	55	48	121	86	60	183	124	79	253	168	95	333	220	112	424	282	131	526	345
	3	24	44	35	38	81	62	49	132	96	62	199	139	82	275	189	97	363	248	114	463	317	134	575	386
8	1	22	40	27	35	72	48	49	114	76	64	176	109	84	243	148	100	320	194	118	408	248	138	507	303
	2	23	44	32	36	80	57	51	128	90	66	195	129	86	269	175	103	356	230	121	454	294	141	564	358
	3	24	47	36	37	87	64	53	139	101	67	210	145	88	290	198	105	384	258	123	492	330	143	612	402
10	1	22	43	28	34	78	50	49	123	78	65	189	113	89	257	154	106	341	200	125	436	257	146	542	314
	2	23	47	33	36	86	59	51	136	93	67	206	134	91	282	182	109	374	238	128	479	305	149	596	372
	3	24	50	37	37	92	67	52	146	104	69	220	150	94	303	205	111	402	268	131	515	342	152	642	417
15	1	21	50	30	33	89	53	47	142	83	64	220	120	88	298	163	110	389	214	134	493	273	162	609	333
	2	22	53	35	35	96	63	49	153	99	66	235	142	91	320	193	112	419	253	137	532	323	165	658	394
	3	24	55	40	36	102	71	51	163	111	68	248	160	93	339	218	115	445	286	140	565	365	167	700	444
20	1	21	54	31	33	99	56	46	157	87	62	246	125	86	334	171	107	436	224	131	552	285	158	681	347
	2	22	57	37	34	105	66	48	167	104	64	259	149	89	354	202	110	463	265	134	587	339	161	725	414
	3	23	60	42	35	110	74	50	176	116	66	271	168	91	371	228	113	486	300	137	618	383	164	764	466
30	1	20	62	33	31	113	59	45	181	93	60	288	134	83	391	182	103	512	238	125	649	305	151	802	372
	2	21	64	39	33	118	70	47	190	110	62	299	158	85	408	215	105	535	282	129	679	360	155	840	439
	3	22	66	44	34	123	79	48	198	124	64	309	178	88	423	242	108	555	317	132	706	405	158	874	494
50	1	19	71	36	30	133	64	43	216	101	57	349	145	78	477	197	97	627	257	120	797	330	144	984	403
	2	21	73	43	32	137	76	45	223	119	59	358	172	81	490	234	100	645	306	123	820	392	148	1,014	478
	3	22	75	48	33	141	86	46	229	134	61	366	194	83	502	263	103	661	343	126	842	441	151	1,043	538
100	1	18	82	37	28	158	66	40	262	104	53	442	150	73	611	204	91	810	266	112	1,038	341	135	1,285	417
	2	19	83	44	30	161	79	42	267	123	55	447	178	75	619	242	94	822	316	115	1,054	405	139	1,306	494
	3	20	84	50	31	163	89	44	272	138	57	452	200	78	627	272	97	834	355	118	1,069	455	142	1,327	555

COMMON VENT CAPACITY

	TYPE B DOUBLE-WALL COMMON VENT DIAMETER (D)																				
	4"			5"			6"			7"			8"			9"			10"		
Vent Height	COMBINED APPLIANCE INPUT RATING IN THOUSANDS OF Btu/h																				
(H) (feet)	FAN +FAN	FAN +NAT	NAT +NAT	FAN +FAN	FAN +NAT	NAT +NAT	FAN +FAN	FAN +NAT	NAT +NAT	FAN +FAN	FAN +NAT	NAT +NAT	FAN +FAN	FAN +NAT	NAT +NAT	FAN +FAN	FAN +NAT	NAT +NAT	FAN +FAN	FAN +NAT	NAT +NAT
6	92	81	65	140	116	103	204	161	147	309	248	200	404	314	260	547	434	335	672	520	410
8	101	90	73	155	129	114	224	178	163	339	275	223	444	348	290	602	480	378	740	577	465
10	110	97	79	169	141	124	243	194	178	367	299	242	477	377	315	649	522	405	800	627	495
15	125	112	91	195	164	144	283	228	206	427	352	280	556	444	365	753	612	465	924	733	565
20	136	123	102	215	183	160	314	255	229	475	394	310	621	499	405	842	688	523	1,035	826	640
30	152	138	118	244	210	185	361	297	266	547	459	360	720	585	470	979	808	605	1,209	975	740
50	167	153	134	279	244	214	421	353	310	641	547	423	854	706	550	1,164	977	705	1,451	1,188	860
100	175	163	NA	311	277	NA	489	421	NA	751	658	479	1,025	873	625	1,408	1,215	800	1,784	1,502	975

(continued)

TABLE 504.3(1)   CHIMNEYS AND VENTS

TABLE 504.3(1)—continued

VENT CONNECTOR CAPACITY

VENT HEIGHT (H) (feet)	CONNECTOR RISE (R) (feet)	12" FAN Min	12" FAN Max	12" NAT Max	14" FAN Min	14" FAN Max	14" NAT Max	16" FAN Min	16" FAN Max	16" NAT Max	18" FAN Min	18" FAN Max	18" NAT Max	20" FAN Min	20" FAN Max	20" NAT Max	22" FAN Min	22" FAN Max	22" NAT Max	24" FAN Min	24" FAN Max	24" NAT Max
		TYPE B DOUBLE-WALL VENT AND DIAMETER (D) — APPLIANCE INPUT RATING LIMITS IN THOUSANDS OF Btu/h																				
6	2	174	764	496	223	1,046	653	281	1,371	853	346	1,772	1,080	NA	NA	NA	NA	NA	NA	NA	NA	NA
6	4	180	897	616	230	1,231	827	287	1,617	1,081	352	2,069	1,370	NA	NA	NA	NA	NA	NA	NA	NA	NA
6	6	NA	NA	NA	NA	NA	NA	NA	NA	NA	NA	NA	NA	NA	NA	NA	NA	NA	NA	NA	NA	NA
8	2	186	822	516	238	1,126	696	298	1,478	910	365	1,920	1,150	NA	NA	NA	NA	NA	NA	NA	NA	NA
8	4	192	952	644	244	1,307	884	305	1,719	1,150	372	2,211	1,460	471	2,737	1,800	560	3,319	2,180	662	3,957	2,590
8	6	198	1,050	772	252	1,445	1,072	313	1,902	1,390	380	2,434	1,770	478	3,018	2,180	568	3,665	2,640	669	4,373	3,130
10	2	196	870	536	249	1,195	730	311	1,570	955	379	2,049	1,205	NA	NA	NA	NA	NA	NA	NA	NA	NA
10	4	201	997	664	256	1,371	924	318	1,804	1,205	387	2,332	1,535	486	2,887	1,890	581	3,502	2,280	686	4,175	2,710
10	6	207	1,095	792	263	1,509	1,118	325	1,989	1,455	395	2,556	1,865	494	3,169	2,290	589	3,849	2,760	694	4,593	3,270
15	2	214	967	568	272	1,334	790	336	1,760	1,030	408	2,317	1,305	NA	NA	NA	NA	NA	NA	NA	NA	NA
15	4	221	1,085	712	279	1,499	1,006	344	1,978	1,320	416	2,579	1,665	523	3,197	2,060	624	3,881	2,490	734	4,631	2,960
15	6	228	1,181	856	286	1,632	1,222	351	2,157	1,610	424	2,796	2,025	533	3,470	2,510	634	4,216	3,030	743	5,035	3,600
20	2	223	1,051	596	291	1,443	840	357	1,911	1,095	430	2,533	1,385	NA	NA	NA	NA	NA	NA	NA	NA	NA
20	4	230	1,162	748	298	1,597	1,064	365	2,116	1,395	438	2,778	1,765	554	3,447	2,180	661	4,190	2,630	772	5,005	3,130
20	6	237	1,253	900	307	1,726	1,288	373	2,287	1,695	450	2,984	2,145	567	3,708	2,650	671	4,511	3,190	785	5,392	3,790
30	2	216	1,217	632	286	1,664	910	367	2,183	1,190	461	2,891	1,540	NA	NA	NA	NA	NA	NA	NA	NA	NA
30	4	223	1,316	792	294	1,802	1,160	376	2,366	1,510	474	3,110	1,920	619	3,840	2,365	728	4,861	2,860	847	5,606	3,410
30	6	231	1,400	952	303	1,920	1,410	384	2,524	1,830	485	3,299	2,340	632	4,080	2,875	741	4,976	3,480	860	5,961	4,150
50	2	206	1,479	689	273	2,023	1,007	350	2,659	1,315	435	3,548	1,665	NA	NA	NA	NA	NA	NA	NA	NA	NA
50	4	213	1,561	860	281	2,139	1,291	359	2,814	1,685	447	3,730	2,135	580	4,601	2,633	709	5,569	3,185	851	6,633	3,790
50	6	221	1,631	1,031	290	2,242	1,575	369	2,951	2,055	461	3,893	2,605	594	4,808	3,208	724	5,826	3,885	867	6,943	4,620
100	2	192	1,923	712	254	2,644	1,050	326	3,490	1,370	402	4,707	1,740	NA	NA	NA	NA	NA	NA	NA	NA	NA
100	4	200	1,984	888	263	2,731	1,346	336	3,606	1,760	414	4,842	2,220	523	5,982	2,750	639	7,254	3,330	769	8,650	3,950
100	6	208	2,035	1,064	272	2,811	1,642	346	3,714	2,150	426	4,968	2,700	539	6,143	3,350	654	7,453	4,070	786	8,892	4,810

COMMON VENT CAPACITY

TYPE B DOUBLE-WALL COMMON VENT DIAMETER (D) — COMBINED APPLIANCE INPUT RATING IN THOUSANDS OF Btu/h

VENT HEIGHT (H) (feet)	12" FAN +FAN	12" FAN +NAT	12" NAT +NAT	14" FAN +FAN	14" FAN +NAT	14" NAT +NAT	16" FAN +FAN	16" FAN +NAT	16" NAT +NAT	18" FAN +FAN	18" FAN +NAT	18" NAT +NAT	20" FAN +FAN	20" FAN +NAT	20" NAT +NAT	22" FAN +FAN	22" FAN +NAT	22" NAT +NAT	24" FAN +FAN	24" FAN +NAT	24" NAT +NAT
6	900	696	588	1,284	990	815	1,735	1,336	1,065	2,253	1,732	1,345	2,838	2,180	1,660	3,488	2,677	1,970	4,206	3,226	2,390
8	994	773	652	1,423	1,103	912	1,927	1,491	1,190	2,507	1,936	1,510	3,162	2,439	1,860	3,890	2,998	2,200	4,695	3,616	2,680
10	1,076	841	712	1,542	1,200	995	2,093	1,625	1,300	2,727	2,113	1,645	3,444	2,665	2,030	4,241	3,278	2,400	5,123	3,957	2,920
15	1,247	986	825	1,794	1,410	1,158	2,440	1,910	1,510	3,184	2,484	1,910	4,026	3,133	2,360	4,971	3,862	2,790	6,016	4,670	3,400
20	1,405	1,116	916	2,006	1,588	1,290	2,722	2,147	1,690	3,561	2,798	2,140	4,548	3,552	2,640	5,573	4,352	3,120	6,749	5,261	3,800
30	1,658	1,327	1,025	2,373	1,892	1,525	3,220	2,558	1,990	4,197	3,326	2,520	5,303	4,193	3,110	6,539	5,157	3,680	7,940	6,247	4,480
50	2,024	1,640	1,280	2,911	2,347	1,863	3,964	3,183	2,430	5,184	4,149	3,075	6,567	5,240	3,800	8,116	6,458	4,500	9,837	7,813	5,475
100	2,569	2,131	1,670	3,732	3,076	2,450	5,125	4,202	3,200	6,749	5,509	4,050	8,597	6,986	5,000	10,681	8,648	5,920	13,004	10,499	7,200

For SI: 1 inch = 25.4 mm, 1 foot = 304.8 mm; British thermal unit per hour = 0.2931 W.

TABLE 504.3(2)
CAPACITY OF TYPE B DOUBLE-WALL VENT WITH SINGLE-WALL CONNECTORS
SERVING TWO OR MORE CATEGORY I APPLIANCES

VENT CONNECTOR CAPACITY

		\																								
		SINGLE-WALL METAL VENT CONNECTOR DIAMETER (D)																								
		3"			4"			5"			6"			7"			8"			9"			10"		
VENT HEIGHT	CONNECTOR RISE	APPLIANCE INPUT RATING IN THOUSANDS OF Btu/h																								
(H)	(R)	FAN		NAT	FAN		NAT	FAN		NAT	FAN		NAT	FAN		NAT	FAN		NAT	FAN		NAT	FAN		NAT
(feet)	(feet)	Min	Max	Max	Min	Max	Max	Min	Max	Max	Min	Max	Max	Min	Max	Max	Min	Max	Max	Min	Max	Max	Min	Max	Max
6	1	NA	NA	26	NA	NA	46	NA	NA	71	NA	NA	102	207	223	140	262	293	183	325	373	234	447	463	286
	2	NA	NA	31	NA	NA	55	NA	NA	85	168	182	123	215	251	167	271	331	219	334	422	281	458	524	344
	3	NA	NA	34	NA	NA	62	121	131	95	175	198	138	222	273	188	279	361	247	344	462	316	468	574	385
8	1	NA	NA	27	NA	NA	48	NA	NA	75	NA	NA	106	226	240	145	285	316	191	352	403	244	481	502	299
	2	NA	NA	32	NA	NA	57	125	126	89	184	193	127	234	266	173	293	353	228	360	450	292	492	560	355
	3	NA	NA	35	NA	NA	64	130	138	100	191	208	144	241	287	197	302	381	256	370	489	328	501	609	400
10	1	NA	NA	28	NA	NA	50	119	121	77	182	186	110	240	253	150	302	335	196	372	429	252	506	534	308
	2	NA	NA	33	84	85	59	124	134	91	189	203	132	248	278	183	311	369	235	381	473	302	517	589	368
	3	NA	NA	36	89	91	67	129	144	102	197	217	148	257	299	203	320	398	265	391	511	339	528	637	413
15	1	NA	NA	29	79	87	52	116	138	81	177	214	116	238	291	158	312	380	208	397	482	266	556	596	324
	2	NA	NA	34	83	94	62	121	150	97	185	230	138	246	314	189	321	411	248	407	522	317	568	646	387
	3	NA	NA	39	87	100	70	127	160	109	193	243	157	255	333	215	331	438	281	418	557	360	579	690	437
20	1	49	56	30	78	97	54	115	152	84	175	238	120	233	325	165	306	425	217	390	538	276	546	664	336
	2	52	59	36	82	103	64	120	163	101	182	252	144	243	346	197	317	453	259	400	574	331	558	709	403
	3	55	62	40	87	107	72	125	172	113	190	264	164	252	363	223	326	476	294	412	607	375	570	750	457
30	1	47	60	31	77	110	57	112	175	89	169	278	129	226	380	175	296	497	230	378	630	294	528	779	358
	2	51	62	37	81	115	67	117	185	106	177	290	152	236	397	208	307	521	274	389	662	349	541	819	425
	3	54	64	42	85	119	76	122	193	120	185	300	172	244	412	235	316	542	309	400	690	394	555	855	482
50	1	46	69	34	75	128	60	109	207	96	162	336	137	217	460	188	284	604	245	364	768	314	507	951	384
	2	49	71	40	79	132	72	114	215	113	170	345	164	226	473	223	294	623	293	376	793	375	520	983	458
	3	52	72	45	83	136	82	119	221	123	178	353	186	235	486	252	304	640	331	387	816	423	535	1,013	518
100	1	45	79	34	71	150	61	104	249	98	153	424	140	205	585	192	269	774	249	345	993	321	476	1,236	393
	2	48	80	41	75	153	73	110	255	115	160	428	167	212	593	228	279	788	299	358	1,011	383	490	1,259	469
	3	51	81	46	79	157	85	114	260	129	168	433	190	222	603	256	289	801	339	368	1,027	431	506	1,280	527

COMMON VENT CAPACITY

	TYPE B DOUBLE-WALL VENT AND DIAMETER (D)																				
	4"			5"			6"			7"			8"			9"			10"		
VENT HEIGHT	COMBINED APPLIANCE INPUT RATING IN THOUSANDS OF Btu/h																				
(H)	FAN	FAN	NAT	FAN	FAN	NAT	FAN	FAN	NAT	FAN	FAN	NAT	FAN	FAN	NAT	FAN	FAN	NAT	FAN	FAN	NAT
(feet)	+FAN	+NAT	+NAT	+FAN	+NAT	+NAT	+FAN	+NAT	+NAT	+FAN	+NAT	+NAT	+FAN	+NAT	+NAT	+FAN	+NAT	+NAT	+FAN	+NAT	+NAT
6	NA	78	64	NA	113	99	200	158	144	304	244	196	398	310	257	541	429	332	665	515	407
8	NA	87	71	NA	126	111	218	173	159	331	269	218	436	342	285	592	473	373	730	569	460
10	NA	94	76	163	137	120	237	189	174	357	292	236	467	369	309	638	512	398	787	617	487
15	121	108	88	189	159	140	275	221	200	416	343	274	544	434	357	738	599	456	905	718	553
20	131	118	98	208	177	156	305	247	223	463	383	302	606	487	395	824	673	512	1,013	808	626
30	145	132	113	236	202	180	350	286	257	533	446	349	703	570	459	958	790	593	1,183	952	723
50	159	145	128	268	233	208	406	337	296	622	529	410	833	686	535	1,139	954	689	1,418	1,157	838
100	166	153	NA	297	263	NA	469	398	NA	726	633	464	999	846	606	1,378	1,185	780	1,741	1,459	948

For SI: 1 inch = 25.4 mm, 1 foot = 304.8 mm, British thermal unit per hour = 0.2931 W.

TABLE 504.3(3)　　　　　　　　　　　　　　　　　　　　　　　　　　　　　　　　　　　CHIMNEYS AND VENTS

TABLE 504.3(3)
CAPACITY OF MASONRY CHIMNEY WITH TYPE B DOUBLE-WALL
CONNECTORS SERVING TWO OR MORE CATEGORY I APPLIANCES

VENT CONNECTOR CAPACITY

VENT HEIGHT (H) (feet)	CONNECTOR RISE (R) (feet)	3" FAN Min	3" FAN Max	3" NAT Max	4" FAN Min	4" FAN Max	4" NAT Max	5" FAN Min	5" FAN Max	5" NAT Max	6" FAN Min	6" FAN Max	6" NAT Max	7" FAN Min	7" FAN Max	7" NAT Max	8" FAN Min	8" FAN Max	8" NAT Max	9" FAN Min	9" FAN Max	9" NAT Max	10" FAN Min	10" FAN Max	10" NAT Max
6	1	24	33	21	39	62	40	52	106	67	65	194	101	87	274	141	104	370	201	124	479	253	145	599	319
	2	26	43	28	41	79	52	53	133	85	67	230	124	89	324	173	107	436	232	127	562	300	148	694	378
	3	27	49	34	42	92	61	55	155	97	69	262	143	91	369	203	109	491	270	129	633	349	151	795	439
8	1	24	39	22	39	72	41	55	117	69	71	213	105	94	304	148	113	414	210	134	539	267	156	682	335
	2	26	47	29	40	87	53	57	140	86	73	246	127	97	350	179	116	473	240	137	615	311	160	776	394
	3	27	52	34	42	97	62	59	159	98	75	269	145	99	383	206	119	517	276	139	672	358	163	848	452
10	1	24	42	22	38	80	42	55	130	71	74	232	108	101	324	153	120	444	216	142	582	277	165	739	348
	2	26	50	29	40	93	54	57	153	87	76	261	129	103	366	184	123	498	247	145	652	321	168	825	407
	3	27	55	35	41	105	63	58	170	100	78	284	148	106	397	209	126	540	281	147	705	366	171	893	463
15	1	24	48	23	38	93	44	54	154	74	72	277	114	100	384	164	125	511	229	153	658	297	184	824	375
	2	25	55	31	39	105	55	56	174	89	74	299	134	103	419	192	128	558	260	156	718	339	187	900	432
	3	26	59	35	41	115	64	57	189	102	76	319	153	105	448	215	131	597	292	159	760	382	190	960	486
20	1	24	52	24	37	102	46	53	172	77	71	313	119	98	437	173	123	584	239	150	752	312	180	943	397
	2	25	58	31	39	114	56	55	190	91	73	335	138	101	467	199	126	625	270	153	805	354	184	1,011	452
	3	26	63	35	40	123	65	57	204	104	75	353	157	104	493	222	129	661	301	156	851	396	187	1,067	505
30	1	24	54	25	37	111	48	52	192	82	69	357	127	96	504	187	119	680	255	145	883	337	175	1,115	432
	2	25	60	32	38	122	58	54	208	95	72	376	145	99	531	209	122	715	287	149	928	378	179	1,171	484
	3	26	64	36	40	131	66	56	221	107	74	392	163	101	554	233	125	746	317	152	968	418	182	1,220	535
50	1	23	51	25	36	116	51	51	209	89	67	405	143	92	582	213	115	798	294	140	1,049	392	168	1,334	506
	2	24	59	32	37	127	61	53	225	102	70	421	161	95	604	235	118	827	326	143	1,085	433	172	1,379	558
	3	26	64	36	39	135	69	55	237	115	72	435	180	98	624	260	121	854	357	147	1,118	474	176	1,421	611
100	1	23	46	24	35	108	50	49	208	92	65	428	155	88	640	237	109	907	334	134	1,222	454	161	1,589	596
	2	24	53	31	37	120	60	51	224	105	67	444	174	92	660	260	113	933	368	138	1,253	497	165	1,626	651
	3	25	59	35	38	130	68	53	237	118	69	458	193	94	679	285	116	956	399	141	1,282	540	169	1,661	705

COMMON VENT CAPACITY

VENT HEIGHT (H) (feet)	12 FAN+FAN	12 FAN+NAT	12 NAT+NAT	19 FAN+FAN	19 FAN+NAT	19 NAT+NAT	28 FAN+FAN	28 FAN+NAT	28 NAT+NAT	38 FAN+FAN	38 FAN+NAT	38 NAT+NAT	50 FAN+FAN	50 FAN+NAT	50 NAT+NAT	63 FAN+FAN	63 FAN+NAT	63 NAT+NAT	78 FAN+FAN	78 FAN+NAT	78 NAT+NAT	113 FAN+FAN	113 FAN+NAT	113 NAT+NAT
6	NA	74	25	NA	119	46	NA	178	71	NA	257	103	NA	351	143	NA	458	188	NA	582	246	1,041	853	NA
8	NA	80	28	NA	130	53	NA	193	82	NA	279	119	NA	384	163	NA	501	218	724	636	278	1,144	937	408
10	NA	84	31	NA	138	56	NA	207	90	NA	299	131	NA	409	177	606	538	236	776	686	302	1,226	1,010	454
15	NA	NA	36	NA	152	67	NA	233	106	NA	334	152	523	467	212	682	611	283	874	781	365	1,374	1,156	546
20	NA	NA	41	NA	NA	75	NA	250	122	NA	368	172	565	508	243	742	668	325	955	858	419	1,513	1,286	648
30	NA	NA	NA	NA	NA	NA	NA	270	137	NA	404	198	615	564	278	816	747	381	1,062	969	496	1,702	1,473	749
50	NA	NA	NA	NA	NA	NA	NA	NA	NA	NA	NA	NA	NA	620	328	879	831	461	1,165	1,089	606	1,905	1,692	922
100	NA	NA	NA	NA	NA	NA	NA	NA	NA	NA	NA	NA	NA	NA	348	NA	NA	499	NA	NA	669	2,053	1,921	1,058

For SI: 1 inch = 25.4 mm, 1 square inch = 645.16 mm^2, 1 foot = 304.8 mm; British thermal unit per hour = 0.2931 W.

TABLE 504.3(4)
CAPACITY OF MASONRY CHIMNEY WITH SINGLE-WALL
CONNECTORS SERVING TWO OR MORE CATEGORY I APPLIANCES

VENT CONNECTOR CAPACITY

		SINGLE-WALL METAL VENT CONNECTOR DIAMETER (D)																								
		3"			**4"**			**5"**			**6"**			**7"**			**8"**			**9"**			**10"**		
VENT HEIGHT	**CONNECTOR RISE**	**APPLIANCE INPUT RATING LIMITS IN THOUSANDS OF Btu/h**																								
(H)	**(R)**	**FAN**		**NAT**	**FAN**		**NAT**	**FAN**		**NAT**	**FAN**		**NAT**	**FAN**		**NAT**	**FAN**		**NAT**	**FAN**		**NAT**	**FAN**		**NAT**
(feet)	**(feet)**	Min	Max	Max	Min	Max	Max	Min	Max	Max	Min	Max	Max	Min	Max	Max	Min	Max	Max	Min	Max	Max	Min	Max	Max
6	1	NA	NA	21	NA	NA	39	NA	NA	66	179	191	100	231	271	140	292	366	200	362	474	252	499	594	316
	2	NA	NA	28	NA	NA	52	NA	NA	84	186	227	123	239	321	172	301	432	231	373	557	299	509	696	376
	3	NA	NA	34	NA	NA	61	134	153	97	193	258	142	247	365	202	309	491	269	381	634	348	519	793	437
8	1	NA	NA	21	NA	NA	40	NA	NA	68	195	208	103	250	298	146	313	407	207	387	530	263	529	672	331
	2	NA	NA	28	NA	NA	52	137	139	85	202	240	125	258	343	177	323	465	238	397	607	309	540	766	391
	3	NA	NA	34	NA	NA	62	143	156	98	210	264	145	266	376	205	332	509	274	407	663	356	551	838	450
10	1	NA	NA	22	NA	NA	41	130	151	70	202	225	106	267	316	151	333	434	213	410	571	273	558	727	343
	2	NA	NA	29	NA	NA	53	136	150	86	210	255	128	276	358	181	343	489	244	420	640	317	569	813	403
	3	NA	NA	34	97	102	62	143	166	99	217	277	147	284	389	207	352	530	279	430	694	363	580	880	459
15	1	NA	NA	23	NA	NA	43	129	151	73	199	271	112	268	376	161	349	502	225	445	646	291	623	808	366
	2	NA	NA	30	92	103	54	135	170	88	207	295	132	277	411	189	359	548	256	456	706	334	634	884	424
	3	NA	NA	34	96	112	63	141	185	101	215	315	151	286	439	213	368	586	289	466	755	378	646	945	479
20	1	NA	NA	23	87	99	45	128	167	76	197	303	117	265	425	169	345	569	235	439	734	306	614	921	387
	2	NA	NA	30	91	111	55	134	185	90	205	325	136	274	455	195	355	610	266	450	787	348	627	986	443
	3	NA	NA	35	96	119	64	140	199	103	213	343	154	282	481	219	365	644	298	461	831	391	639	1,042	496
30	1	NA	NA	24	86	108	47	126	187	80	193	347	124	259	492	183	338	665	250	430	864	330	600	1,089	421
	2	NA	NA	31	91	119	57	132	203	93	201	366	142	269	518	205	348	699	282	442	908	372	613	1,145	473
	3	NA	NA	35	95	127	65	138	216	105	209	381	160	277	540	229	358	729	312	452	946	412	626	1,193	524
50	1	NA	NA	24	85	113	50	124	204	87	188	392	139	252	567	208	328	778	287	417	1,022	383	582	1,302	492
	2	NA	NA	31	89	123	60	130	218	100	196	408	158	262	588	230	339	806	320	429	1,058	425	596	1,346	545
	3	NA	NA	35	94	131	68	136	231	112	205	422	176	271	607	255	349	831	351	440	1,090	466	610	1,386	597
100	1	NA	NA	23	84	104	49	122	200	89	182	410	151	243	617	232	315	875	328	402	1,181	444	560	1,537	580
	2	NA	NA	30	88	115	59	127	215	102	190	425	169	253	636	254	326	899	361	415	1,210	488	575	1,570	634
	3	NA	NA	34	93	124	67	133	228	115	199	438	188	262	654	279	337	921	392	427	1,238	529	589	1,604	687

COMMON VENT CAPACITY

	MINIMUM INTERNAL AREA OF MASONRY CHIMNEY FLUE (square inches)																							
	12			**19**			**28**			**38**			**50**			**63**			**78**			**113**		
VENT HEIGHT	**COMBINED APPLIANCE INPUT RATING IN THOUSANDS OF Btu/h**																							
(H)	**FAN +FAN**	**FAN +NAT**	**NAT +NAT**	**FAN +FAN**	**FAN +NAT**	**NAT +NAT**	**FAN +FAN**	**FAN +NAT**	**NAT +NAT**	**FAN +FAN**	**FAN +NAT**	**NAT +NAT**	**FAN +FAN**	**FAN +NAT**	**NAT +NAT**	**FAN +FAN**	**FAN +NAT**	**NAT +NAT**	**FAN +FAN**	**FAN +NAT**	**NAT +NAT**	**FAN +FAN**	**FAN +NAT**	**NAT +NAT**
(feet)																									
6	NA	NA	25	NA	118	45	NA	176	71	NA	255	102	NA	348	142	NA	455	187	NA	579	245	NA	846	NA
8	NA	NA	28	NA	128	52	NA	190	81	NA	276	118	NA	380	162	NA	497	217	NA	633	277	1,136	928	405
10	NA	NA	31	NA	136	56	NA	205	89	NA	295	129	NA	405	175	NA	532	234	771	680	300	1,216	1,000	450
15	NA	NA	36	NA	NA	66	NA	230	105	NA	335	150	NA	400	210	677	602	280	866	772	360	1,359	1,139	540
20	NA	NA	NA	NA	NA	74	NA	247	120	NA	362	170	NA	503	240	765	661	321	947	849	415	1,495	1,264	640
30	NA	NA	NA	NA	NA	NA	NA	NA	135	NA	398	195	NA	558	275	808	739	377	1,052	957	490	1,682	1,447	740
50	NA	NA	NA	NA	NA	NA	NA	NA	NA	NA	NA	NA	NA	612	325	NA	821	456	1,152	1,076	600	1,879	1,672	910
100	NA	NA	NA	NA	NA	NA	NA	NA	NA	NA	NA	NA	NA	NA	NA	NA	NA	494	NA	NA	663	2,006	1,885	1,046

For SI: 1 inch= 25.4 mm, 1 square inch = 645.16 mm^2, 1 foot = 304.8 mm, 1 British thermal unit per hour = 0.2931 W.

TABLE 504.3(5)

TABLE 504.3(5)
CAPACITY OF SINGLE-WALL METAL PIPE OR TYPE B ASBESTOS CEMENT
VENT SERVING TWO OR MORE DRAFT HOOD-EQUIPPED APPLIANCES

VENT CONNECTOR CAPACITY

TOTAL VENT HEIGHT (H) (feet)	CONNECTOR RISE (R) (feet)	VENT CONNECTOR DIAMETER (D)					
		3"	4"	5"	6"	7"	8"
		MAXIMUM APPLIANCE INPUT RATING IN THOUSANDS OF Btu/h					
6 - 8	1	21	40	68	102	146	205
	2	28	53	86	124	178	235
	3	34	61	98	147	204	275
15	1	23	44	77	117	179	240
	2	30	56	92	134	194	265
	3	35	64	102	155	216	298
30 and up	1	25	49	84	129	190	270
	2	31	58	97	145	211	295
	3	36	68	107	164	232	321

COMMON VENT CAPACITY

TOTAL VENT HEIGHT (H) (feet)	COMMON VENT DIAMETER						
	4"	5"	6"	7"	8"	10"	12"
	COMBINED APPLIANCE INPUT RATING IN THOUSANDS OF Btu/h						
6	48	78	111	155	205	320	NA
8	55	89	128	175	234	365	505
10	59	95	136	190	250	395	560
15	71	115	168	228	305	480	690
20	80	129	186	260	340	550	790
30	NA	147	215	300	400	650	940
50	NA	NA	NA	360	490	810	1,190

For SI: 1 inch= 25.4 mm, 1 foot = 304.8 mm, 1 British thermal unit per hour = 0.2931 W.

TABLE 504.3(6)
EXTERIOR MASONRY CHIMNEY, SINGLE NAT INSTALLATIONS WITH TYPE B DOUBLE-WALL VENT CONNECTORS

VENT HEIGHT (feet)	MINIMUM ALLOWABLE INPUT RATING OF SPACE-HEATING APPLIANCE IN THOUSANDS OF Btu/h							
	Internal area of chimney (square inches)							
	12	19	28	38	50	63	78	113
37°F or Greater	Local 99% Winter Design Temperature: 37°F or Greater							
6	0	0	0	0	0	0	0	0
8	0	0	0	0	0	0	0	0
10	0	0	0	0	0	0	0	0
15	NA	0	0	0	0	0	0	0
20	NA	NA	123	190	249	184	0	0
30	NA	NA	NA	NA	NA	393	334	0
50	NA	NA	NA	NA	NA	NA	NA	579
27 to 36°F	Local 99% Winter Design Temperature: 27 to 36°F							
6	0	0	68	116	156	180	212	266
8	0	0	82	127	167	187	214	263
10	0	51	97	141	183	201	225	265
15	NA	NA	NA	NA	233	253	274	305
20	NA	NA	NA	NA	NA	307	330	362
30	NA	NA	NA	NA	NA	419	445	485
50	NA	NA	NA	NA	NA	NA	NA	763
17 to 26°F	Local 99% Winter Design Temperature: 17 to 26°F							
6	NA	NA	NA	NA	NA	215	259	349
8	NA	NA	NA	NA	197	226	264	352
10	NA	NA	NA	NA	214	245	278	358
15	NA	NA	NA	NA	NA	296	331	398
20	NA	NA	NA	NA	NA	352	387	457
30	NA	NA	NA	NA	NA	NA	507	581
50	NA	NA	NA	NA	NA	NA	NA	NA
5 to 16°F	Local 99% Winter Design Temperature: 5 to 16°F							
6	NA	NA	NA	NA	NA	NA	NA	416
8	NA	NA	NA	NA	NA	NA	312	423
10	NA	NA	NA	NA	NA	289	331	430
15	NA	NA	NA	NA	NA	NA	393	485
20	NA	NA	NA	NA	NA	NA	450	547
30	NA	NA	NA	NA	NA	NA	NA	682
50	NA	NA	NA	NA	NA	NA	NA	972
-10 to 4°F	Local 99% Winter Design Temperature: -10 to 4°F							
6	NA	NA	NA	NA	NA	NA	NA	484
8	NA	NA	NA	NA	NA	NA	NA	494
10	NA	NA	NA	NA	NA	NA	NA	513
15	NA	NA	NA	NA	NA	NA	NA	586
20	NA	NA	NA	NA	NA	NA	NA	650
30	NA	NA	NA	NA	NA	NA	NA	805
50	NA	NA	NA	NA	NA	NA	NA	1,003
-11°F or Lower	Local 99% Winter Design Temperature: -11°F or Lower							
	Not recommended for any vent configurations							

For SI: °C = [(°F) - 32]/1.8, 1 inch = 25.4 mm, 1 foot = 304.8 mm, 1 British thermal unit per hour = 0.2931 W.

For local 99% Winter Design Temperature refer to "Winter Design Dry-bulb Temperature" column in Table 302.1 of the *Energy Conservation Construction Code of New York State.*

TABLE 504.3(7)　　　　　　　　　　　　　　　　　　　　　　　CHIMNEYS AND VENTS

TABLE 504.3(7)
EXTERIOR MASONRY CHIMNEY, NAT + NAT INSTALLATIONS WITH
TYPE B DOUBLE-WALL VENT CONNECTORS

504.3(7a) Combined Appliance Maximum Input Rating in Thousands of Btu per Hour

VENT HEIGHT (feet)	Internal Area of Chimney (square inches)							
	12	19	28	38	50	63	78	113
6	25	46	71	103	143	188	246	NA
8	28	53	82	119	163	218	278	408
10	31	56	90	131	177	236	302	454
15	NA	67	106	152	212	283	365	546
20	NA	NA	NA	NA	NA	325	419	648
30	NA	NA	NA	NA	NA	NA	496	749
50	NA	NA	NA	NA	NA	NA	NA	922
100	NA	NA	NA	NA	NA	NA	NA	NA

504.3(7b) Minimum Allowable Input Rating of Space-Heating Appliance in Thousands of Btu per Hour

VENT HEIGHT (feet)	Internal Area of Chimney (square inches)							
	12	19	28	38	50	63	78	113
37°F or Greater	Local 99% Winter Design Temperature: 37°F or Greater							
6	0	0	0	0	0	0	0	NA
8	0	0	0	0	0	0	0	0
10	0	0	0	0	0	0	0	0
15	NA	0	0	0	0	0	0	0
20	NA	NA	NA	NA	NA	184	0	0
30	NA	NA	NA	NA	NA	393	334	0
50	NA	NA	NA	NA	NA	NA	NA	579
100	NA	NA	NA	NA	NA	NA	NA	NA
27 to 36°F	Local 99% Winter Design Temperature: 27 to 36°F							
6	0	0	68	NA	NA	180	212	NA
8	0	0	82	NA	NA	187	214	263
10	0	51	NA	NA	NA	201	225	265
15	NA	NA	NA	NA	NA	253	274	305
20	NA	NA	NA	NA	NA	307	330	362
30	NA	NA	NA	NA	NA	NA	445	485
50	NA	NA	NA	NA	NA	NA	NA	763
100	NA	NA	NA	NA	NA	NA	NA	NA
17 to 26°F	Local 99% Winter Design Temperature: 17 to 26°F							
6	NA	NA	NA	NA	NA	NA	NA	NA
8	NA	NA	NA	NA	NA	NA	264	352
10	NA	NA	NA	NA	NA	NA	278	358
15	NA	NA	NA	NA	NA	NA	331	398
20	NA	NA	NA	NA	NA	NA	387	457
30	NA	NA	NA	NA	NA	NA	NA	581
50	NA	NA	NA	NA	NA	NA	NA	862
100	NA	NA	NA	NA	NA	NA	NA	NA
5 to 16°F	Local 99% Winter Design Temperature: 5 to 16°F							
6	NA	NA	NA	NA	NA	NA	NA	NA
8	NA	NA	NA	NA	NA	NA	NA	NA
10	NA	NA	NA	NA	NA	NA	NA	430
15	NA	NA	NA	NA	NA	NA	NA	485
20	NA	NA	NA	NA	NA	NA	NA	547
30	NA	NA	NA	NA	NA	NA	NA	682
50	NA	NA	NA	NA	NA	NA	NA	NA
100	NA	NA	NA	NA	NA	NA	NA	NA
4°F or Lower	Local 99% Winter Design Temperature: 4°F or Lower							
	Not recommended for any vent configurations							

For SI: °C = [(°F) - 32]/1.8, 1 inch = 25.4 mm, 1 square inch = 645.16 mm², 1 foot = 304.8 mm, 1 British thermal unit per hour = 0.2931 W.

N
Y For local 99% Winter Design Temperature refer to "Winter Design Dry-bulb Temperature" column in Table 302.1 of the *Energy Conservation Construction*
N
Y *Code of New York State.*

TABLE 504.3(8)
EXTERIOR MASONRY CHIMNEY, FAN + NAT INSTALLATIONS WITH
TYPE B DOUBLE-WALL VENT CONNECTORS

504.3(8a) Combined Appliance Maximum Input Rating in Thousands of Btu per Hour

VENT HEIGHT (feet)	Internal Area of Chimney (square inches)							
	12	19	28	38	50	63	78	113
6	74	119	178	257	351	458	582	853
8	80	130	193	279	384	501	636	937
10	84	138	207	299	409	538	686	1010
15	NA	152	233	334	467	611	781	1156
20	NA	NA	250	368	508	668	858	1286
30	NA	NA	NA	404	564	747	969	1473
50	NA	NA	NA	NA	NA	831	1089	1692
100	NA	NA	NA	NA	NA	NA	NA	1921

504.3(8b) Minimum Allowable Input Rating of Space-Heating Appliance in Thousands of Btu per Hour

VENT HEIGHT (feet)	Internal Area of Chimney (square inches)							
	12	19	28	38	50	63	78	113
37°F or Greater	Local 99% Winter Design Temperature: 37°F or Greater							
6	0	0	0	0	0	0	0	0
8	0	0	0	0	0	0	0	0
10	0	0	0	0	0	0	0	0
15	NA	0	0	0	0	0	0	0
20	NA	NA	123	190	249	184	0	0
30	NA	NA	NA	334	398	393	334	0
50	NA	NA	NA	NA	NA	714	707	579
100	NA	NA	NA	NA	NA	NA	NA	1600
27 to 36°F	Local 99% Winter Design Temperature: 27 to 36°F							
6	0	0	68	116	156	180	212	266
8	0	0	82	127	167	187	214	263
10	0	51	97	141	183	210	225	265
15	NA	111	142	183	233	253	274	305
20	NA	NA	187	230	284	307	330	362
30	NA	NA	NA	330	319	419	445	485
50	NA	NA	NA	NA	NA	672	705	763
100	NA	NA	NA	NA	NA	NA	NA	1554
17 to 26°F	Local 99% Winter Design Temperature: 17 to 26°F							
6	0	55	99	141	182	215	259	349
8	52	74	111	154	197	226	264	352
10	NA	90	125	169	214	245	278	358
15	NA	NA	167	212	263	296	331	398
20	NA	NA	212	258	316	352	387	457
30	NA	NA	NA	362	429	470	507	581
50	NA	NA	NA	NA	NA	723	766	862
100	NA	NA	NA	NA	NA	NA	NA	1669
5 to 16°F	Local 99% Winter Design Temperature: 5 to 16°F							
6	NA	78	121	166	214	252	301	416
8	NA	94	135	182	230	269	312	423
10	NA	111	149	198	250	289	331	430
15	NA	NA	193	247	305	346	393	485
20	NA	NA	NA	293	360	408	450	547
30	NA	NA	NA	377	450	531	580	682
50	NA	NA	NA	NA	NA	797	853	972
100	NA	NA	NA	NA	NA	NA	NA	1833
-10 to 4°F	Local 99% Winter Design Temperature: -10 to 4°F							
6	NA	NA	145	196	249	296	349	484
8	NA	NA	159	213	269	320	371	494
10	NA	NA	175	231	292	339	397	513
15	NA	NA	NA	283	351	404	457	586
20	NA	NA	NA	333	408	468	528	650
30	NA	NA	NA	NA	NA	603	667	805
50	NA	NA	NA	NA	NA	NA	955	1003
100	NA	NA	NA	NA	NA	NA	NA	NA
-11°F or Lower	Local 99% Winter Design Temperature: -11°F or Lower							
	Not recommended for any vent configurations							

For SI: °C = [(°F) - 32]/1.8, 1 inch = 25.4 mm; 1 square inch = 645.16 mm², 1 foot = 304.8 mm, 1 British thermal unit per hour = 0.2931 W.

For local 99% Winter Design Temperature refer to "Winter Design Dry-bulb Temperature" column in Table 302.1 of the *Energy Conservation Construction Code of New York State.*

N
Y
N
Y

CHAPTER 6
SPECIFIC APPLIANCES

SECTION 601
GENERAL

601.1 Scope. This chapter shall govern the approval, design, installation, construction, maintenance, alteration and repair of the appliances and equipment specifically identified herein.

601.2 Fireplaces. Fireplaces (solid fuel type or ANSI Z21.50) shall be installed with tight fitting noncombustible fireplace doors to control infiltration losses in the construction types listed here:
1. Masonry or factory-built fireplaces designed to allow an open burn.
2. Whenever a decorative appliance (ANSI Standard Z21.60 gas-log style unit) is installed in a vented solid fuel fireplace.
3. Vented decorative gas fireplace appliances ANSI Standard Z21.50 unit.

Fireplaces shall be provided with a source of combustion air, as required by the fireplace construction provisions of the *Building Code of New York State* or *Residential Code of New York State*.

601.3 Flame safeguard device. All fuel gas space heating appliances installed or used in a building occupied as a residence shall be equipped with an automatic flame safeguard device that shall shut off the fuel supply to the main burner or group of burners when the flame or pilot light thereof is extinguished.

SECTION 602
DECORATIVE APPLIANCES
FOR INSTALLATION IN FIREPLACES

602.1 General. Decorative appliances for installation in approved solid fuel burning fireplaces shall be tested in accordance with ANSI Z21.60 and shall be installed in accordance with the manufacturer's installation instructions. Manually lighted natural gas decorative appliances shall be tested in accordance with ANSI Z21.84.

602.2 Flame safeguard device. Decorative appliances for installation in approved solid fuel-burning fireplaces, with the exception of those tested in accordance with ANSI Z21.84, shall utilize a direct ignition device, an ignitor or a pilot flame to ignite the fuel at the main burner, and shall be equipped with a flame safeguard device. The flame safeguard device shall automatically shut off the fuel supply to a main burner or group of burners when the means of ignition of such burners becomes inoperative.

602.3 Prohibited installations. Decorative appliances for installation in fireplaces shall not be installed where prohibited by Section 303.3.

SECTION 603
LOG LIGHTERS

603.1 General. Log lighters shall be tested in accordance with IAS 8 and shall be installed in accordance with the manufacturer's installation instructions.

603.2 Automatic valves. Automatic valves or semi-automatic valves shall be provided and shall comply with the applicable provisions of ANSI Z21.21.

SECTION 604
VENTED GAS FIREPLACES
(DECORATIVE APPLIANCES)
AND VENTED GAS FIREPLACE HEATERS

604.1 General. Vented gas fireplaces shall be tested in accordance with ANSI Z21.50, shall be installed in accordance with the manufacturer's installation instructions and shall be designed and equipped as specified in Section 602.2.

604.2 Access. Panels, grilles, and access doors that are required to be removed for normal servicing operations shall not be attached to the building.

604.3 Vented gas fireplace heaters–general. Vented gas fireplace heaters shall be installed in accordance with the manufacturer's installation instructions, shall be tested in accordance with ANSI Z21.88 and shall be designed and equipped as specified in Section 602.2.

SECTION 605
INCINERATORS AND CREMATORIES

605.1 General. Incinerators and crematories shall be installed in accordance with the manufacturer's installation instructions.

SECTION 606
COMMERCIAL-INDUSTRIAL INCINERATORS

606.1 Incinerators, commercial-industrial. Commercial-industrial type incinerators shall be constructed and installed in accordance with NFPA 82.

SECTION 607
VENTED WALL FURNACES

607.1 General. Vented wall furnaces shall be tested in accordance with ANSI Z21.49 or Z21.86/CSA 2.32 and shall be installed in accordance with the manufacturer's installation instructions.

607.2 Venting. Vented wall furnaces shall be vented in accordance with Section 503.

607.3 Location. Vented wall furnaces shall be located so as not to cause a fire hazard to walls, floors, combustible furnishings or doors. Vented wall furnaces installed between bathrooms and adjoining rooms shall not circulate air from bathrooms to other parts of the building.

607.4 Door swing. Vented wall furnaces shall be located so that a door cannot swing within 12 inches (305 mm) of an air inlet or air outlet of such furnace measured at right angles to the opening. Doorstops or door closers shall not be installed to obtain this clearance.

607.5 Ducts prohibited. Ducts shall not be attached to wall furnaces. Casing extension boots shall not be installed unless listed as part of the appliance.

607.6 Access. Vented wall furnaces shall be provided with access for cleaning of heating surfaces, removal of burners, replacement of sections, motors, controls, filters and other working parts, and for adjustments and lubrication of parts requiring such attention. Panels, grilles and access doors that are required to be removed for normal servicing operations shall not be attached to the building construction.

SECTION 608
FLOOR FURNACES

608.1 General. Floor furnaces shall be tested in accordance with ANSI Z21.48 or Z21.86/CSA 2.32 and shall be installed in accordance with the manufacturer's installation instructions.

608.2 Placement. The following provisions apply to floor furnaces.
1. Floors. Floor furnaces shall not be installed in the floor of any doorway, stairway landing, aisle, or passageway of any enclosure, public or private, or in an exitway from any such room or space.
2. Walls and corners. The register of a floor furnace with a horizontal warm-air outlet shall not be placed closer than 6 inches (152 mm) to the nearest wall. A distance of at least 18 inches (457 mm) from two adjoining sides of the floor furnace register to walls shall be provided to eliminate the necessity of occupants walking over the warm-air discharge. The remaining sides shall be permitted to be placed not closer than 6 inches (152 mm) to a wall. Wall-register models shall not be placed closer than 6 inches (152 mm) to a corner.
3. Draperies. The furnace shall be placed so that a door, drapery, or similar object cannot be nearer than 12 inches (305 mm) to any portion of the register of the furnace.
4. Floor construction. Floor furnaces shall not be installed in concrete floor construction built on grade.

5. Thermostat. The controlling thermostat for a floor furnace shall be located within the same room or space as the floor furnace or shall be located in an adjacent room or space that is permanently open to the room or space containing the floor furnace.

608.3 Bracing. The floor around the furnace shall be braced and headed with a support framework designed in accordance with the *Building Code of New York State*.

608.4 Clearance. The lowest portion of the floor furnace shall have not less than a 6-inch (152 mm) clearance from the grade level; except where the lower 6-inch (152 mm) portion of the floor furnace is sealed by the manufacturer to prevent entrance of water, the minimum clearance shall be not less than 2 inches (51 mm). Where such clearances cannot be provided, the ground below and to the sides shall be excavated to form a pit under the furnace so that the required clearance is provided beneath the lowest portion of the furnace. A 12-inch (305 mm) minimum clearance shall be provided on all sides except the control side, which shall have an 18-inch (457 mm) minimum clearance.

608.5 First floor installation. Where the basement story level below the floor in which a floor furnace is installed is utilized as habitable space, such floor furnaces shall be enclosed as specified in Section 608.6 and shall project into a nonhabitable space.

608.6 Upper floor installations. Floor furnaces installed in upper stories of buildings shall project below into nonhabitable space and shall be separated from the nonhabitable space by an enclosure constructed of noncombustible materials. The floor furnace shall be provided with access, clearance to all sides and bottom of not less than 6 inches and combustion air in accordance with Section 304.

SECTION 609
DUCT FURNACES

609.1 General. Duct furnaces shall be tested in accordance with ANSI Z83.9 or UL 795 and shall be installed in accordance with the manufacturer's installation instructions.

609.2 Access panels. Ducts connected to duct furnaces shall have removable access panels on both the upstream and downstream sides of the furnace.

609.3 Location of draft hood and controls. The controls, combustion air inlets, and draft hoods for duct furnaces shall be located outside of the ducts. The draft hood shall be located in the same enclosure from which combustion air is taken.

609.4 Circulating air. Where a duct furnace is installed so that supply ducts convey air to areas outside the space containing the furnace, the return air shall also be conveyed by a duct(s) sealed to the furnace casing and terminating outside the space containing the furnace.

The duct furnace shall be installed on the positive pressure side of the circulating air blower.

SECTION 610
DIRECT-FIRED MAKE-UP AIR HEATERS

610.1 General. Direct-fired make-up air heaters shall be tested in accordance with ANSI Z83.4 and shall be installed in accordance with the manufacturer's installation instructions.

610.2 Installation. Direct-fired make-up air heaters shall not be used to supply any area containing sleeping quarters.

610.3 Outdoor air. All air handled by a direct-fired make-up air heater, including combustion air, shall be brought in from outdoors.

> **Exception:** Indoor air added to the outdoor airstream after the outdoor airstream has passed the combustion zone.

610.4 Outdoor air louvers. If outdoor air louvers of either the manual or automatic type are used, such devices shall be proved in the open position prior to allowing the main burners to operate.

610.5 Controls. Direct-fired make-up air heaters shall be equipped with airflow-sensing devices, safety shutoff devices, operating temperature controls, and thermally actuated temperature limit controls in accordance with the terms of their listings.

610.6 Atmospheric vents and gas reliefs or bleeds. Direct-fired make-up air heaters with valve train components equipped with atmospheric vents or gas reliefs or bleeds shall have their atmospheric vent lines or gas reliefs or bleeds lead to the outdoors. Means shall be employed on these lines to prevent water from entering and to prevent blockage by insects and foreign matter. An atmospheric vent line shall not be required to be provided on a valve train component equipped with a listed vent limiter.

610.7 Relief opening. The design of the installation shall include provision to permit direct-fired make-up air heaters to operate at rated capacity by taking into account the structure's designed infiltration rate, providing properly designed relief openings or an interlocked power exhaust system, or a combination of these methods. The structure's designed infiltration rate and the size of relief openings shall be determined by approved engineering methods. Relief openings shall be permitted to be louvers or counterbalanced gravity dampers. Motorized dampers or closable louvers shall be permitted to be used, provided they are verified to be in their full open position prior to main burner operation.

SECTION 611
DIRECT-FIRED INDUSTRIAL AIR HEATERS

611.1 General. Direct-fired industrial air heaters shall be tested in accordance with ANSI Z83.18 and shall be installed in accordance with the manufacturer's installation instructions.

611.2 Location. Direct-fired industrial air heaters shall be installed only in industrial and commercial occupancies. Direct-fired air heaters shall not be installed in any area intended for sleeping. Direct-fired heaters shall not be installed in hazardous locations where room air is recirculated across the burner or which contain substances that are made toxic by exposure to flames.

611.3 Installation. Direct-fired industrial air heaters shall be permitted to be installed in accordance with their listing and the manufacturer's instructions. Direct-fired industrial air heaters shall be installed only in industrial or commercial occupancies. Direct-fired industrial air heaters shall be permitted to provide fresh air ventilation.

611.4 Clearance from combustible materials. Direct-fired industrial air heaters shall be installed with a clearance from combustible material of not less than that shown on the label and in the manufacturers' instructions.

611.5 Air supply. Air to direct-fired industrial air heaters shall be taken from the building, ducted directly from outdoors, or a combination of both. Direct-fired industrial air heaters shall incorporate a means to supply outside ventilation air to the space at a rate of not less than 4 cfm per 1,000 Btu per hour (0.38 m^3 per min per Kw) of rated input of the heater. If a separate means is used to supply ventilation air, an interlock shall be provided so as to lock out the main burner operation until the mechanical means is verified. If outside air dampers or closing louvers are used, they shall be verified to be in the open position prior to main burner operation.

611.6 Atmospheric vents, gas reliefs, or bleeds. Direct-fired industrial air heaters with valve train components equipped with atmospheric vents, gas reliefs, or bleeds shall have their atmospheric vent lines and gas reliefs or bleeds lead to the outdoors.

Means shall be employed on these lines to prevent water from entering and to prevent blockage by insects and foreign matter. An atmospheric vent line shall not be required to be provided on a valve train component equipped with a listed vent limiter.

611.7 Relief opening. The design of the installation shall include adequate provision to permit direct-fired industrial air heaters to operate at rated capacity by taking into account the structure's designed infiltration rate, providing properly designed relief openings or an interlocked power exhaust system, or a combination of these methods. The structure's designed infiltration rate and the size of relief openings shall be determined by approved engineering methods. Relief openings shall be permitted to be louvers or counterbalanced gravity dampers. Motorized dampers or closable louvers shall be permitted to be used, provided they are verified to be in their full open position prior to main burner operation.

SECTION 612
CLOTHES DRYERS

612.1 General. Clothes dryers shall be tested in accordance with ANSI Z21.5.1 or ANSI Z21.5.2 and shall be installed in accordance with the manufacturer's installation instructions.

SECTION 613
CLOTHES DRYER EXHAUST

613.1 Installation. Clothes dryers shall be exhausted in accordance with the manufacturer's instructions. Dryer exhaust systems shall be independent of all other systems and shall convey the moisture and any products of combustion to the outside of the building.

613.2 Duct penetrations. Ducts that exhaust clothes dryers shall not penetrate or be located within any fireblocking, draftstopping or any wall, floor/ceiling or other assembly required by the *Building Code of New York State* to be fire-resistance rated, unless such duct is constructed of galvanized steel or aluminum of the thickness specified in Table 603.3 of the *Mechanical Code of New York State* and the fire-resistance rating is maintained in accordance with the *Building Code of New York State*.

613.3 Cleaning access. Each vertical duct riser for dryers listed to ANSI Z21.5.2 shall be provided with a cleanout or other means for cleaning the interior of the duct.

613.4 Exhaust material. Exhaust ducts for clothes dryers shall terminate on the outside of the building and shall be equipped with a backdraft damper. Screens shall not be installed at the duct termination. Ducts shall not be connected or installed with sheet metal screws or other fasteners that will obstruct the flow. Clothes dryer exhaust ducts shall not be connected to a vent connector, vent or chimney. Clothes dryer exhaust ducts shall not extend into or through ducts or plenums.

613.5 Makeup air. Installations exhausting more than 200 cfm (0.09 m³/s) shall be provided with makeup air. Where a closet is designed for the installation of a clothes dryer, an opening having an area of not less than 100 square inches (645 mm²) for makeup air shall be provided in the closet enclosure, or makeup air shall be provided by other approved means.

613.6 Domestic clothes dryer ducts. Exhaust ducts for domestic clothes dryers shall have a smooth interior finish. The exhaust duct shall be a minimum nominal size of 4 inches (102 mm) in diameter. The entire exhaust system shall be supported and secured in place. The male end of the duct at overlapped duct joints shall extend in the direction of airflow. Clothes dryer transition ducts used to connect the appliance to the exhaust duct system shall be metal and limited to a single length not to exceed 8 feet (2438 mm) in length and shall be listed and labeled for the application. Transition ducts shall not be concealed within construction.

613.6.1 Maximum length. The maximum length of a clothes dryer exhaust duct shall not exceed 25 feet (7620 mm) from the dryer location to the outlet terminal. The maximum length of the duct shall be reduced 2¹/₂ feet (762 mm) for each 45-degree (0.79 rad) bend and 5 feet (1524 mm) for each 90-degree (1.6 rad) bend.

Exception: Where the make and model of the clothes dryer to be installed is known and the manufacturer's installation instructions for such dryer are provided, the maximum length of the exhaust duct, including any transition duct, shall be permitted to be in accordance with the dryer manufacturer's installation instructions.

613.6.2 Rough-in required. Where a compartment or space for a domestic clothes dryer is provided, an exhaust duct system shall be installed.

613.7 Commercial clothes dryers. The installation of dryer exhaust ducts serving Type 2 clothes dryers shall comply with the appliance manufacturer's installation instructions. Exhaust fan motors installed in exhaust systems shall be located outside of the airstream. In multiple installations, the fan shall operate continuously or be interlocked to operate when any individual unit is operating. Ducts shall have a minimum clearance of 6 inches (152 mm) to combustible materials.

SECTION 614
SAUNA HEATERS

614.1 General. Sauna heaters shall be installed in accordance with the manufacturer's installation instructions.

614.2 Location and protection. Sauna heaters shall be located so as to minimize the possibility of accidental contact by a person in the room.

614.2.1 Guards. Sauna heaters shall be protected from accidental contact by an approved guard or barrier of material having a low coefficient of thermal conductivity. The guard shall not substantially affect the transfer of heat from the heater to the room.

614.3 Access. Panels, grilles and access doors that are required to be removed for normal servicing operations shall not be attached to the building.

614.4 Combustion and dilution air intakes. Sauna heaters of other than the direct-vent type shall be installed with the draft hood and combustion air intake located outside the sauna room. Where the combustion air inlet and the draft hood are in a dressing room adjacent to the sauna room, there shall be provisions to prevent physically blocking the combustion air inlet and the draft hood inlet, and to prevent physical contact with the draft hood and vent assembly, or warning notices shall be posted to avoid such contact. Any warning notice shall be easily readable, shall contrast with its background, and the wording shall be in letters not less than 1/4 inch (6.4 mm) high.

614.5 Combustion and ventilation air. Combustion air shall not be taken from inside the sauna room. Combustion and ventilation air for a sauna heater not of the direct-vent type shall be provided to the area in which the combustion air inlet and draft hood are located in accordance with Section 304.

614.6 Heat and time controls. Sauna heaters shall be equipped with a thermostat which will limit room temperature to 194°F (90°C). If the thermostat is not an integral part of the sauna heater, the heat-sensing element shall be located within 6 inches (152 mm) of the ceiling. If the heat-sensing element is a capillary tube and bulb, the assembly shall be attached to the wall or other support, and shall be protected against physical damage.

614.6.1 Timers. A timer, if provided to control main burner operation, shall have a maximum operating time of 1 hour. The control for the timer shall be located outside the sauna room.

614.7 Sauna room. A ventilation opening into the sauna room shall be provided. The opening shall be not less than 4 inches by 8 inches (102 mm by 203 mm) located near the top of the door into the sauna room.

614.7.1 Warning notice. The following permanent notice, constructed of approved material, shall be mechanically attached to the sauna room on the outside:

> WARNING: DO NOT EXCEED 30 MINUTES IN SAUNA. EXCESSIVE EXPOSURE CAN BE HARMFUL TO HEALTH. ANY PERSON WITH POOR HEALTH SHOULD CONSULT A PHYSICIAN BEFORE USING SAUNA.

The words shall contrast with the background and the wording shall be in letters not less than 1/4 inch (6.4 mm) high.

Exception: This section shall not apply to one- and two-family dwellings.

SECTION 615
ENGINE AND GAS TURBINE-POWERED EQUIPMENT

615.1 Powered equipment. Permanently installed equipment powered by internal combustion engines and turbines shall be installed in accordance with the manufacturer's installation instructions and NFPA 37.

SECTION 616
POOL AND SPA HEATERS

616.1 General. Pool and spa heaters shall be tested in accordance with ANSI Z21.56 and shall be installed in accordance with the manufacturer's installation instructions.

SECTION 617
FORCED-AIR WARM-AIR FURNACES

617.1 General. Forced-air warm-air furnaces shall be tested in accordance with ANSI Z21.47 or UL 795 and shall be installed in accordance with the manufacturer's installation instructions.

617.2 Forced-air furnaces. The minimum unobstructed total area of the outside and return air ducts or openings to a forced-air warm-air furnace shall be not less than 2 square inches for each 1,000 Btu/h (4402 mm^2/W) output rating capacity of the furnace and not less than that specified in the furnace manufacturer's installation instructions. The minimum unobstructed total area of supply ducts from a forced-air warm-air furnace shall be not less than 2 square inches for each 1,000 Btu/h (4402 mm^2/W) output rating capacity of the furnace and not less than that specified in the furnace manufacturer's installation instructions.

Exception: The total area of the supply air ducts and outside and return air ducts shall not be required to be larger than the minimum size required by the furnace manufacturer's installation instructions.

617.3 Dampers. Volume dampers shall not be placed in the air inlet to a furnace in a manner which will reduce the required air to the furnace.

617.4 Circulating air ducts for forced-air warm-air furnaces. Circulating air for fuel-burning, forced-air-type, warm-air furnaces shall be conducted into the blower housing from outside the furnace enclosure by continuous airtight ducts.

617.5 Prohibited sources. Outside or return air for a forced-air heating system shall not be taken from the following locations:

1. Closer than 10 feet (3048 mm) from an appliance vent outlet, a vent opening from a plumbing drainage system or the discharge outlet of an exhaust fan, unless the outlet is 3 feet (914 mm) above the outside air inlet.

2. Where there is the presence of objectionable odors, fumes or flammable vapors; or where located less than 10 feet (3048 mm) above the surface of any abutting public way or driveway; or where located at grade level by a sidewalk, street, alley or driveway.

3. A hazardous or insanitary location or a refrigeration machinery room as defined in the *Mechanical Code of New York State*.

4. A room or space, the volume of which is less than 25 percent of the entire volume served by such system. Where connected by a permanent opening having an area sized in accordance with Section 617.2, adjoining rooms or spaces shall be considered as a single room or space for the purpose of determining the volume of such rooms or spaces.

> **Exception:** The minimum volume requirement shall not apply where the amount of return air taken from a room or space is less than or equal to the amount of supply air delivered to such room or space.

5. A room or space containing an appliance where such a room or space serves as the sole source of return air.

> **Exception:** This shall not apply where:
>
> 1. The appliance is a direct-vent appliance or an appliance not requiring a vent in accordance with Section 501.8.
> 2. The room or space complies with the following requirements:
>   2.1. The return air shall be taken from a room or space having a volume exceeding 1 cubic foot for each 10 Btu/h (9.6 L/W) of combined input rating of all fuel-burning appliances therein.
>   2.2. The volume of supply air discharged back into the same space shall be approximately equal to the volume of return air taken from the space.
>   2.3. Return-air inlets shall not be located within 10 feet (3048 mm) of any appliance firebox or draft hood in the same room or space.
> 3. Rooms or spaces containing solid fuel-burning appliances, provided that return-air inlets are located not less than 10 feet (3048 mm) from the firebox of such appliances.

6. A closet, bathroom, toilet room, kitchen, garage, mechanical room, boiler room or furnace room.

617.6 Screen. Required outdoor air inlets for residential portions of a building shall be covered with a screen having 1/4 inch (6.4 mm) openings. Required outdoor air inlets serving a nonresidential portion of a building shall be covered with screen having openings larger than 1/4 inch (6.4 mm) and not larger than 1 inch (25 mm).

617.7 Return-air limitation. Return air from one dwelling unit shall not be discharged into another dwelling unit.

SECTION 618
CONVERSION BURNERS

618.1 Conversion burners. The installation of conversion burners shall conform to ANSI Z21.8.

SECTION 619
UNIT HEATERS

619.1 General. Unit heaters shall be tested in accordance with ANSI Z83.8 and shall be installed in accordance with the manufacturer's installation instructions.

619.2 Support. Suspended-type unit heaters shall be supported by elements that are designed and constructed to accommodate the weight and dynamic loads. Hangers and brackets shall be of noncombustible material.

619.3 Ductwork. Ducts shall not be connected to a unit heater unless the heater is listed for such installation.

619.4 Clearance. Suspended-type unit heaters shall be installed with clearances to combustible materials of not less than 18 inches (457 mm) at the sides, 12 inches (305 mm) at the bottom and 6 inches (152 mm) above the top where the unit heater has an internal draft hood or 1 inch (25 mm) above the top of the sloping side of the vertical draft hood.

Floor-mounted-type unit heaters shall be installed with clearances to combustible materials at the back and one side only of not less than 6 inches (152 mm). Where the flue gases are vented horizontally, the 6-inch (152 mm) clearance shall be measured from the draft hood of vent instead of the rear wall of the unit heater. Floor-mounted-type unit heaters shall not be installed on combustible floors unless listed for such installation.

Clearances for servicing all unit heaters shall be in accordance with the manufacturer's installation instructions.

> **Exception:** Unit heaters listed for reduced clearance shall be permitted to be installed with such clearances in accordance with their listing and the manufacturer's instructions.

SECTION 620
UNVENTED ROOM HEATERS

620.1 General. Unvented room heaters shall be tested in accordance with ANSI Z21.11.2 and shall be installed in accordance with the conditions of the listing and the manufacturer's installation instructions. Unvented room heaters utilizing fuels other than fuel gas shall be regulated by the *Mechanical Code of New York State*.

620.2 Prohibited use. One or more unvented room heaters shall not be used as the sole source of comfort heating in a dwelling unit.

620.3 Input rating. Unvented room heaters shall not have an input rating in excess of 40,000 Btu/h (11.7 kW).

620.4 Prohibited locations. Unvented room heaters shall not be installed within occupancies in Use Groups A, E and I. The location of unvented room heaters shall also comply with Section 303.3.

620.5 Room or space volume. The aggregate input rating of all unvented appliances installed in a room or space shall not exceed 20 Btu/h per cubic foot (207 W/m^3) of volume of such room or space. Where the room or space in which the equipment is installed is directly connected to another room or space by a doorway, archway or other opening of comparable size that cannot be closed, the volume of such adjacent room or space shall be permitted to be included in the calculations.

620.6 Oxygen-depletion safety system. Unvented room heaters shall be equipped with an oxygen-depletion-sensitive safety shutoff system. The system shall shut off the gas supply to the main and pilot burners when the oxygen in the surrounding atmosphere is depleted to the percent concentration specified by the manufacturer, but not lower than 18 percent. The system shall not incorporate field adjustment means capable of changing the set point at which the system acts to shut off the gas supply to the room heater.

620.7 Unvented log heaters. An unvented log heater shall not be installed in a factory-built fireplace unless the fireplace system has been specifically tested, listed and labeled for such use in accordance with UL 127.

SECTION 621
VENTED ROOM HEATERS

621.1 General. Vented room heaters shall be tested in accordance with ANSI Z21.11.1 or Z21.86/CSA 2.32, shall be designed and equipped as specified in Section 602.2 and shall be installed in accordance with the manufacturer's installation instructions.

SECTION 622
COOKING APPLIANCES

622.1 Cooking appliances. Cooking appliances that are designed for permanent installation, including ranges, ovens, stoves, broilers, grills, fryers, griddles, hot plates and barbecues, shall be tested in accordance with ANSI Z21.1, ANSI Z21.58, or ANSI Z83.11 and shall be installed in accordance with the manufacturer's installation instructions.

622.2 Prohibited location. Cooking appliances designed, tested, listed and labeled for use in commercial occupancies shall not be installed within dwelling units or within any area where domestic cooking operations occur.

622.3 Domestic appliances. Cooking appliances installed within dwelling units and within areas where domestic cooking operations occur shall be listed and labeled as household-type appliances for domestic use.

622.4 Domestic range installation. Domestic ranges installed on combustible floors shall be set on their own bases or legs and shall be installed with clearances of not less than that shown on the label.

622.5 Open top broiler unit hoods. A ventilating hood shall be provided above a domestic open-top broiler unit, unless otherwise listed for forced down draft ventilation.

622.5.1 Clearances. A minimum clearance of 24 inches (610 mm) shall be maintained between the cooking top and combustible material above the hood. The hood shall be at least as wide as the open top broiler unit and be centered over the unit.

622.6 Commercial cooking appliance venting. Commercial cooking appliances, other than those exempted by Section 501.8, shall be vented by connecting the appliance to a vent or chimney in accordance with this code and the appliance manufacturer's instructions or the appliance shall be vented in accordance with Section 501.1.1.

SECTION 623
WATER HEATERS

623.1 General. Water heaters shall be tested in accordance with ANSI Z21.10.1 and ANSI Z21.10.3 and shall be installed in accordance with the manufacturer's installation instructions. Water heaters utilizing fuels other than fuel gas shall be regulated by the *Mechanical Code of New York State*.

623.1.1 Installation requirements. The requirements for water heaters relative to sizing, relief valves, drain pans and scald protection shall be in accordance with the *Plumbing Code of New York State*.

623.2 Water heaters utilized for space heating. Water heaters utilized both to supply potable hot water and provide hot water for space-heating applications shall be listed and labeled for such applications by the manufacturer and shall be installed in accordance with the manufacturer's installation instructions and the *Plumbing Code of New York State*.

SECTION 624
REFRIGERATORS

624.1 General. Refrigerators shall be tested in accordance with ANSI Z21.19 and shall be installed in accordance with the manufacturer's installation instructions.

Refrigerators shall be provided with adequate clearances for ventilation at the top and back, and shall be installed in accordance with the manufacturer's instructions. If such instructions are not available, at least 2 inches (51 mm) shall be provided between the back of the refrigerator and the wall and at least 12 inches (305 mm) above the top.

SECTION 625
GAS-FIRED TOILETS

625.1 General. Gas-fired toilets shall be tested in accordance with ANSI Z21.61 and shall be installed in accordance with the manufacturer's installation instructions.

625.2 Clearance. A gas-fired toilet shall be installed in accordance with its listing and the manufacturer's instructions, provided that the clearance shall in any case be sufficient to afford ready access for use, cleanout and necessary servicing.

SECTION 626
AIR-CONDITIONING EQUIPMENT

626.1 General. Gas-fired air-conditioning equipment shall be tested in accordance with ANSI Z21.40.1 or ANSI Z21.40.2 and shall be installed in accordance with the manufacturer's installation instructions.

626.2 Independent piping. Gas piping serving heating equipment shall be permitted also to serve cooling equipment where such heating and cooling equipment cannot be operated simultaneously (see Section 402).

626.3 Connection of gas engine-powered air conditioners. To protect against the effects of normal vibration in service, gas engines shall not be rigidly connected to the gas supply piping.

626.4 Clearances for indoor installation. Air-conditioning equipment installed in rooms other than alcoves and closets shall be installed with clearances not less than those specified in Section 308.3, except that air-conditioning equipment listed for installation at lesser clearances than those specified in Section 308.3, shall be permitted to be installed in accordance with such listing and the manufacturer's instructions and air-conditioning equipment listed for installation at greater clearances than those specified in Section 308.3, shall be installed in accordance with such listing and the manufacturer's instructions.

Air-conditioning equipment installed in rooms other than alcoves and closets shall be permitted to be installed with reduced clearances to combustible material, provided that the combustible material is protected in accordance with Table 308.2.

626.5 Alcove and closet installation. Air-conditioning equipment installed in spaces such as alcoves and closets shall be specifically listed for such installation and installed in accordance with the terms of such listing. The installation clearances for air-conditioning equipment in alcoves and closets shall not be reduced by the protection methods described in Table 308.2.

626.6 Installation. Air-conditioning equipment shall be installed in accordance with the manufacturer's instructions. Unless the equipment is listed for installation on a combustible surface such as a floor or roof, or unless the surface is protected in an approved manner, equipment shall be installed on a surface of noncombustible construction with noncombustible material and surface finish and with no combustible material against the underside thereof.

626.7 Plenums and air ducts. A plenum supplied as a part of the air-conditioning equipment shall be installed in accordance with the equipment manufacturer's instructions. Where a plenum is not supplied with the equipment, such plenum shall be installed in accordance with the fabrication and installation instructions provided by the plenum and equipment manufacturer. The method of connecting supply and return ducts shall facilitate proper circulation of air.

Where air-conditioning equipment is installed within a space separated from the spaces served by the equipment, the air circulated by the equipment shall be conveyed by ducts that are sealed to the casing of the equipment and that separate the circulating air from the combustion and ventilation air.

626.8 Refrigeration coils. A refrigeration coil shall not be installed in conjunction with a forced-air furnace where circulation of cooled air is provided by the furnace blower, unless the blower has sufficient capacity to overcome the external static resistance imposed by the duct system and cooling coil at the air throughput necessary for heating or cooling, whichever is greater. Furnaces shall not be located upstream from cooling units, unless the cooling unit is designed or equipped so as not to develop excessive temperature or pressure. Refrigeration coils shall be installed in parallel with or on the downstream side of central furnaces to avoid condensation in the heating element, unless the furnace has been specifically listed for downstream installation. With a parallel flow arrangement, the dampers or other means used to control flow of air shall be sufficiently tight to prevent any circulation of cooled air through the furnace.

Means shall be provided for disposal of condensate and to prevent dripping of condensate onto the heating element.

626.9 Cooling units used with heating boilers. Boilers, where used in conjunction with refrigeration systems, shall be installed so that the chilled medium is piped in parallel with the heating boiler with appropriate valves to prevent the chilled medium from entering the heating boiler. Where hot water heating boilers are connected to heating coils located in air handling units where they may be exposed to refrigerated air circulation, such boiler piping systems shall be equipped with flow control valves or other automatic means to prevent gravity circulation of the boiler water during the cooling cycle.

626.10 Switches in electrical supply line. Means for interrupting the electrical supply to the air-conditioning equipment and to its associated cooling tower (if supplied and installed in a location remote from the air conditioner) shall be provided within sight of and not over 50 feet (15 240 mm) from the air conditioner and cooling tower.

SECTION 627
ILLUMINATING APPLIANCES

627.1 General. Illuminating appliances shall be tested in accordance with ANSI Z21.42 and shall be installed in accordance with the manufacturer's installation instructions.

627.2 Mounting on buildings. Illuminating appliances designed for wall or ceiling mounting shall be securely attached to substantial structures in such a manner that they are not dependent on the gas piping for support.

627.3 Mounting on posts. Illuminating appliances designed for post mounting shall be securely and rigidly attached to a post. Posts shall be rigidly mounted. The strength and rigidity of posts greater than 3 feet (914 mm) in height shall be at least equivalent to that of a $2 \frac{1}{2}$-inch diameter (64 mm) post constructed of 0.064-inch (1.6-mm) thick steel or a 1-inch (25.4 mm) Schedule 40 steel pipe. Posts 3 feet (914 mm) or less in height shall not be smaller than a $3/4$-inch (19.1 mm) Schedule 40 steel pipe. Drain openings shall be provided near the base of posts where there is a possibility of water collecting inside them.

627.4 Appliance pressure regulators. Where an appliance pressure regulator is not supplied with an illuminating appliance and the service line is not equipped with a service pressure regulator, an appliance pressure regulator shall be installed in the line to the illuminating appliance. For multi-

ple installations, one regulator of adequate capacity shall be permitted to serve more than one illuminating appliance.

SECTION 628
SMALL CERAMIC KILNS

628.1 General. Ceramic kilns with a maximum interior volume of 20 cubic feet (0.566 m^3) and used for hobby and noncommercial purposes shall be installed in accordance with the manufacturer's installation instructions and the provisions of this code.

SECTION 629
INFRARED RADIANT HEATERS

629.1 General. Infrared radiant heaters shall be tested in accordance with ANSI Z83.6 and shall be installed in accordance with the manufacturer's installation instructions.

629.2 Support. Infrared radiant heaters shall be safely and adequately fixed in an approved position independent of gas and electric supply lines. Hanger and brackets shall be of noncombustible material.

SECTION 630
BOILERS

630.1 Standards. Boilers shall be listed in accordance with the requirements of ANSI Z21.13 or UL 795. The boiler shall be designed and constructed in accordance with the requirements of ASME CSD-1 and as applicable, the ASME Boiler and Pressure Vessel Code I, II, IV, V and IX, NFPA 8501, NFPA 8502, and NFPA 8504. Low pressure boilers shall conform to the requirements of the New York State Department of Labor, 13 NYCRR, Industrial Code Rule 4, and high pressure boilers shall conform to the requirements of the New York State Department of Labor, 13 NYCRR, Industrial Code Rule 14.

630.2 Installation. In addition to the requirements of this code, the installation of boilers shall be in accordance with the manufacturer's instructions and the *Mechanical Code of New York State.* Operating instructions of a permanent type shall be attached to the boiler. Boilers shall have all controls set, adjusted and tested by the installer. A complete control diagram together with complete boiler operating instructions shall be furnished by the installer. The manufacturer's rating data and the nameplate shall be attached to the boiler.

630.3 Clearance to combustible materials. Clearances to combustible materials shall be in accordance with Section 308.4.

SECTION 631
EQUIPMENT INSTALLED IN EXISTING UNLISTED BOILERS

631.1 General. Gas equipment installed in existing unlisted boilers shall comply with Section 630.1 and shall be installed in accordance with the manufacturer's instructions and the *Mechanical Code of New York State*.

SECTION 632
CHIMNEY DAMPER OPENING AREA

632.1 Free opening area of chimney dampers. Where an unlisted decorative appliance for installation in a vented fireplace is installed, the fireplace damper shall have a permanent free opening equal to or greater than specified in Table 632.1.

SECTION 633
FUEL CELL POWER PLANTS

633.1 General. Stationary fuel cell power plants shall be tested in accordance with ANSI Z21.83 and shall be installed in accordance with the manufacturer's installation instructions.

TABLE 632.1
FREE OPENING AREA OF CHIMNEY DAMPER FOR VENTING FLUE GASES
FROM UNLISTED DECORATIVE APPLIANCES FOR INSTALLATION IN VENTED FIREPLACES

CHIMNEY HEIGHT (feet)	MINIMUM PERMANENT FREE OPENING (square inches)[a]						
	8	13	20	29	39	51	64
	Appliance input rating (Btu per hour)						
6	7,800	14,000	23,200	34,000	46,400	62,400	80,000
8	8,400	15,200	25,200	37,000	50,400	68,000	86,000
10	9,000	16,800	27,600	40,400	55,800	74,400	96,400
15	9,800	18,200	30,200	44,600	62,400	84,000	108,800
20	10,600	20,200	32,600	50,400	68,400	94,000	122,200
30	11,200	21,600	36,600	55,200	76,800	105,800	138,600

For SI: 1 foot = 304.8 mm, 1 square inch = 645.16 m², 1 British thermal unit per hour = 0.2931 W.

a. The first six minimum permanent free openings (8 to 51 square inches) correspond approximately to the cross-sectional areas of chimneys having diameters of 3 through 8 inches, respectively. The 64-square-inch opening corresponds to the cross-sectional area of standard 8 inch by 8 inch chimney tile.

CHAPTER 7
REFERENCED STANDARDS

ANSI

American National Standards Institute
11 West 42nd Street
New York, NY 10036

Standard Reference Number	Title	Referenced in Code Section Number
LC 1-97	Gas Piping Systems Using Corrugate Stainless Steel Tubing	403.5.4
Z21.1-96	Household Cooking Gas Appliances	622.1
Z21.5.1-99	Gas Clothes Dryers - Volume I - Type 1 Clothes Dryers	612.1
Z21.5.2-99	Gas Clothes Dryers - Volume II - Type 2 Clothes Dryers with Z21.5.2a-99 and Z21.5.2b-99 Addenda	612.1, 613.3
Z21.8-94	Installation of Domestic Gas Conversion Burners	618.1
Z21.10.1-98	Gas Water Heaters - Volume I Storage, Water Heaters with Input Ratings of 75,000 Btu per Hour or Less	623.1
Z21.10.3-98	Gas Water Heaters - Volume III - Storage, Water Heaters with Input Ratings Above 75,000 Btu per hour, Circulating and Instantaneous Water Heaters - with Z21.10.3a-99 Addendum	623.1
Z21.11.1-91	Gas-Fired Room Heaters - Volume I - Vented Room Heaters - with 1993 Addendum (Replaced by Z21.86-98/CSA 2.32 - M98, Vented Gas-Fired Space Heating Appliances)	621.1
Z21.11.2-96	Gas-Fired Room Heaters - Volume II - Unvented Room Heaters - with Z21.11.2a-97 and Z21.11.2b-98 Addenda	620.1
Z21.13-99	Gas-Fired Low-Pressure Steam and Hot Water Boilers - with Z21.13a-1993 and Z21.13b-1994 Addenda	630.1
Z21.15-97	Manually Operated Gas Valves for Appliances, Appliance Connector Valves, and Hose End Valves	409.1.1

ANSI–continued

Z21.19-90 (R 1999)	Refrigerators Using Gas . . . . .624.1 Fuel - with Z721.19a-1992 (R1999) and Z21.19b-1995 (R1999) Addenda	
Z21.21-95	Standard for Automatic Valves for Gas Appliances 603.2	
Z21.40.1-96	Gas-Fired Absorption Summer . . . . .626.1 Air-Conditioning Appliances - with Z21.40.1a-98 Addendum	
Z21.40.2-96	Gas-Fired Work Activated . . . . .626.1 Air-Conditioning and Heat Pump Appliances (Internal Combustion) - with Z21.40.2a-97 Addendum	
Z21.42-93	Gas-Fired Illuminating Appliances 627.1	
Z21.47-98	Gas-Fired Central Furnaces - . . . . .617.1 with Z21.47a-97 Addendum	
Z21.48-92	Gas-Fired Gravity and Fan-Type . . . . .608.1 Floor Furnaces - with 1993 Addendum (Replaced by Z21.86-98/CSA 2.32-M98, Vented Gas-Fired Space Heating Appliances)	
Z21.49-92	Gas-Fired Gravity and Fan-Type . . . . .607.1 Vented Wall Furnaces - with 1993 Addendum (Replaced by Z21.86-98/CSA 2.32-M98, Vented Gas-Fired Space Heating Appliances)	
Z21.50-98	Vented Gas Fireplaces 604.1	
Z21.56-98	Gas-Fired Pool Heaters - . . . . .616.1 with Z21.56a-99 Addenda	
Z21.58-95	Outdoor Cooking Gas Appliances - . . . . .622.1 with Z21.58a-1998 Addendum	
Z21.60-96	Decorative Gas Appliances for . . . . .602.1 Installation in Solid-Fuel-Burning Fireplaces	
Z21.61-83 (R 1996)	Gas-Fired Toilets 625.1	
Z21.69-97	Connectors for Movable Gas Appliances 411.1	
Z21.83-98	Fuel Cell Power Plants 633.1	
Z21.84-99	Manually-Lighted, Decorative Gas Appliances for Installation in Solid-Fuel-Burning Fireplaces 602.1, 602.2	
Z21.86-98/CSA 2.32-M98	Gas-Fired Vented Space Heating Appliances 607.1, 608.1, 621.1	
Z21.88-99	Vented Gas Fireplace Heaters 604.3	
Z83.4-99	Direct Gas-Fired Make-Up Air Heaters 610.1	

Z83.6-90	Gas-Fired Infrared Heaters	.629.1
Z83.8-96	Gas Unit Heaters - with Z83.8a-1997 Addendum	.619.1
Z83.9-90	Gas-Fired Duct Furnaces - with Z83.9a-1992 Addendum	.609.1
Z83.11-96	Gas Food Service Equipment (Ranges and Unit Broilers), Baking and Roasting Ovens, Fat Fryers, Counter Appliances and Kettles, Steam Cookers, and Steam Generators - with Z83.11a-1997 Addendum	.622.1
Z83.18-90	Direct Gas-Fired Industrial Air Heaters - with Z83.18a-1991 and Z83.18b-1992 Addenda	.611.1

ASME

American Society of Mechanical Engineers
Three Park Avenue
New York, NY 10016-5990

Standard Reference Number	Title	Referenced in Code Section Number
B1.20.1-1983 (R92)	Pipe Threads, General Purpose (inch)	.403.9
B16.1-98	Cast Iron Pipe Flanges and Flanged Fittings, Class 25, 125, 250, and 800	.403.12
B16.20-98	Metallic Gaskets for Pipe Flanges: Ring-Joint, Spiral-Wound, and Jacketed	.403.12
B16.33-90	Manually Operated Metallic Gas Valves for Use in Gas Piping Systems up to 125 psig (Sizes $1/2$ through 2)	.409.1.1
B36.10M-96	Welded and Seamless Wrought - Steel Pipe	.403.4.2
ASME-98	ASME Boiler & Pressure Vessel Code with the 1999 Addendum and Vol. 45 Interpretations	.630.1
CSD-1-98	Controls and Safety Devices for Automatically Fired Boilers with the ASME CSD-1a-1999 Addendum	.630.1

ASTM

American Society for Testing and Materials
100 Barr Harbor Drive
West Conshohocken, PA 19428-2959

Standard Reference Number	Title	Referenced in Code Section Number
A 53- 99b	Specification for Pipe, Steel, Black and Hot-Dipped Zinc-Coated Welded and Seamless	.403.4.2

ASTM–continued

A 106-99[E01]   Specification for Seamless 403.4.2
                      Carbon Steel Pipe for High-
                      Temperature Service

A 254-97   Specification for Copper 403.5.1
                      Brazed Steel Tubing

A 539-99   Specification for Electric 403.5.1
                      Resistance-Welded Coiled Steel Tubing for
                      Gas and Fuel Oil Lines

B 88-99   Specification for Seamless 403.5.2
                      Copper Water Tube

B 210-95   Specification for Aluminum 403.5.3
                      and Aluminum-Alloy Drawn Seamless Tubes

B 241-99   Specification for Aluminum 403.4.4,
                      and Aluminum-Alloy, Seamless   403.5.3
                      Pipe and Seamless Extruded Tube

B 280-99   Specification for Seamless 403.5.2
                      Copper Tube for Air Conditioning and Refrigeration Field Service

C 64-94   Specification for Refractories for Incinerators and Boilers 503.10.2.5

C 315-00   Specification for Clay Flue Linings 501.12

D 2513-99A   Specification for Thermoplastic 403.6, 403.6.1,
                      Gas Pressure Pipe, Tubing, and Fittings   403.11, 404.14.2

E 814-97   Test Method for Fire Tests of 202, 302.2.2.1.2,
                      Through-penetration Fire Stops   302.2.3.1.2

AWWA

American Water Works Association
6666 West Quincy Avenue
Denver, CO 80235

Standard Reference Number	Title	Referenced in Code Section Number
C111-95	Rubber-Gasket Joints . . . . .403.12 A21.11-95 for Ductile-Iron Pressure Pipe and Fittings	

DOTn

Department of Transportation
General Service Administration
7th & D Streets Specification Section,
Room 6039
Washington, DC 20407

Standard Reference Number	Title	Referenced in Code Section Number
49 CFR, Parts 192.281(e) & 192.283 (b)	Transportation of Natural and Other Gas by Pipeline: Minimum Federal Safety Standards 403.6.1	

IAS

International Approval Services
8501 East Pleasant Valley Road
Cleveland, OH 44131

Standard Reference Number	Title	Referenced in Code Section Number
IAS 8-93	Requirements for Gas-Fired . . . . .603.1 Log Lighters for Wood Burning Fireplaces	

ICBO

International Conference of Building Officials
5360 Workman Mill Road
Whittier, CA 90601-2298

Standard Reference Number	Title	Referenced in Code Section Number
BC-NYS-2002	Building Code of New York State 201.3, 301.10, 301.11, 301.12, 301.14, 306.3.1, 306.4.1, 306.5.2, 306.6, 309.2, 401.1.1, 412.6, 413.3, 413.3.1, 413.8.2.4, 501.1, 501.3, 501.12, 501.15.4, 608.3, 613.2	
ECCC-NYS-2002	Energy Conservation Construction Code of New York State 301.2	
FC-NYS-2002	Fire Code of New York State 201.3, 303.4, 401.2, 412.1, 412.6, 412.7, 412.7.3, 412.8, 413.1, 413.3, 413.3.1 413.4, 413.8.2.5	
MC-NYS-2002	Mechanical Code of New York State 101.2.1, 201.3, 301.1.1, 301.13, 304.15, 501.1, 613.2, 617.5, 620.1, 623.1, 630.2, 631.1	
PC-NYS-2002	Plumbing Code of New York State 201.3, 301.6, 623.1.1, 623.2	
PMC-NYS-2002	Property Maintenance Code of New York State 201.3	

MSS

Manufacturers Standardization Society of
the Valve and Fittings Industry
127 Park Street, Northeast
Vienna, VA 22180

Standard Reference Number	Title	Referenced in Code Section Number
SP-6-96	Standard Finishes for Contact Faces of Pipe Flanges and Connecting End Flanges of Valves and Fittings	403.12
SP-58-93	Pipe Hangers and Supports - Materials, Design and Manufacture	407.2

NFPA

National Fire Protection Association
1 Batterymarch Park
P.O. Box 9101
Quincy, MA 02269-9101

Standard Reference Number	Title	Referenced in Code Section Number
37-98	Stationary Combustion Engines and Gas Turbines	615.1
51-97	Design and Installation of Oxygen-Fuel Gas Systems for Welding, Cutting, and Allied Processes	414.1
58- 98	Liquified Petroleum Gases Code	401.2, 402.5.1, 403.6.2, 403.11
82-99	Incinerators, Waste and Linen Handling Systems and Equipment	606.1
88B-97	Repair Garages	305.3
211-00	Chimneys, Fireplaces, Vents, and Solid-Fuel-Burning Appliances	503.5.2, 503.5.3 503.5.6.1, 503.5.6.3
8501-97	Single Burner Boiler Operation	630.1
8502-99	Prevention of Furnace Explosions/Implosions in Multiple Burner Boiler	630.1
8504-96	Atmospheric Fluidized-Bed Boiler Operation	630.1

UL

Underwriters Laboratories Inc.
333 Pfingsten Road
Northbrook, IL 60062

Standard Reference Number	Title	Referenced in Code Section Number
103-95	Factory-Built Chimneys, Residential Type and Building Heating Appliance - with Revisions thru February 1996	506.1
127-96	Factory-Built Fireplaces with Revisions thru January 1998	620.7
441-96	Gas Vents - with Revisions thru October 1997	502.1
641-95	Low Temperature Venting Systems, Type L	502.1
795-95	Commercial-Industrial Gas Heating Equipment - with Revisions thru January 1996	609.1, 617.1, 630.1
959-95	Medium Heat Appliance Factory Built Chimneys - with Revisions thru April 15, 1998	506.3
1738-93	Venting Systems for Gas-Burning Appliances, Categories II, III and IV	502.1
1777-96	Chimney Liners - with Revisions thru August, 1998	501.12, 501.15.4

APPENDIX A
SIZING AND CAPACITIES OF GAS PIPING

(This appendix is informative and is not part of the code.)

In order to determine the size of piping to be used in designing a gas piping system, the following factors must be considered:

1. Allowable loss in pressure from point of delivery to equipment.
2. Maximum gas demand.
3. Length of piping and number of fittings.
4. Specific gravity of the gas.
5. Diversity factor.

For any gas piping system, for special gas utilization equipment, or for conditions other than those covered by Tables 402.3(1) through 402.3(4), or Tables 402.3(15), 402.3(16), or 402.3(17) such as longer runs, greater gas demands, or greater pressure drops, the size of each gas piping system should be determined by standard engineering practices acceptable to the authority having jurisdiction.

Description of tables

(a) The quantity of gas to be provided at each outlet should be determined, whenever possible, directly from the manufacturer's Btu input rating of the equipment which will be installed. In case the ratings of the equipment to be installed are not known, Table A-1 shows the approximate consumption of average appliances of certain types in Btu per hour.

  To obtain the cubic feet per hour of gas required, divide the total Btu input of all equipment by the average Btu heating value per cubic foot of the gas. The average Btu per cubic foot of the gas in the area of the installation may be obtained from the serving gas supplier.

(b) Capacities for gas at low pressure [0.5 psig (35 kPa gauge) or less] in cubic feet per hour of 0.60 specific gravity gas for different sizes and lengths are shown in Tables 402.3(1) and 402.3(2) for iron pipe or equivalent rigid pipe, in Tables 402.3(3) and 402.3(4) for smooth wall semirigid tubing, and Tables 402.3(18), 402.3(19), and 402.3(20) for corrugated stainless steel tubing. Tables 402.3(1) and 402.3(3) are based upon a pressure drop of 0.3 inch (75 pa) water column, whereas Tables 402.3(2), 402.3(4), and 402.3(18) are based upon a pressure drop of 0.5 inch (125 pa) water column. Tables 402.3(19) and 402.3(20) are special low-pressure applications based upon pressure drops

greater than 0.5 inch water column (125 pa). In using these tables, no additional allowance is necessary for an ordinary number of fittings.

(c) Capacities in thousands of Btu per hour of undiluted liquefied petroleum gases based on a pressure drop of 0.5 inch (125 pa) water column for different sizes and lengths are shown in Table 402.3(14) for iron pipe or equivalent rigid pipe and in Table 402.3(15) for smooth wall semirigid tubing, and in Table 402.3(23) for corrugated stainless steel tubing. Tables 402.3(24) and 402.3(25) for corrugated stainless steel tubing are based on pressure drops greater than 0.5 inches water column (125 pa). In using these tables, no additional allowance is necessary for an ordinary number of fittings.

(d) Gas piping systems that are to be supplied with gas of a specific gravity of 0.70 or less can be sized directly from Tables 402.3(1) through 402.3(4), unless the authority having jurisdiction specifies that a gravity factor be applied. Where the specific gravity of the gas is greater than 0.70, the gravity factor should be applied.

  Application of the gravity factor converts the figures given in Tables 402.3(1) through 402.3(4) to capacities with another gas of different specific gravity. Such application is accomplished by multiplying the capacities given in Tables 402.3(1) through 402.3(4) by the multipliers shown in Table 402.3(13). In case the exact specific gravity does not appear in the table, choose the next higher value specific gravity shown.

(e) Capacities for gas at pressures greater than 0.5 psig (3.5 kPa gauge) in cubic feet per hour of 0.60 specific gravity gas for different sizes and lengths are shown in Tables 402.3(5) to 402.3(12) for iron pipe or equivalent rigid pipe and Tables 402.3(23) and 402.3(24) for corrugated stainless steel tubing.

Use of capacity tables

To determine the size of each section of gas piping in a system within the range of the capacity tables, proceed as follows (also see sample calculation at end of Appendix A):

1. Determine the gas demand of each appliance to be attached to the piping system. Where Tables 402.3(1) through 402.3(23) are to be used to select the piping size, calculate the gas demand in terms of cubic feet per hour for each piping system outlet.

Where Tables 402.3(25) through 402.3(34) are to be used to select the piping size, calculate the gas demand in terms of thousands of Btu per hour for each piping system outlet.

2. Where the piping system is for use with other than undiluted liquefied petroleum gases, determine the design system pressure, the allowable loss in pressure (pressure drop), and the specific gravity of the gas to be used in the piping system.

3. Measure the length of piping from the point of delivery to the most remote outlet in the building. Where a multipressure gas piping system is used, gas piping shall be sized for the maximum length of pipe measured from the gas pressure regulator to the most remote outlet of each similarly pressured section.

4. In the appropriate capacity table, select the column showing the measured length, or the next longer length if the table does not give the exact length. This is the only length used in determining the size of any section of gas piping. If the gravity factor is to be applied, the values in the selected column of the table are multiplied by the appropriate multiplier from Table 402.3(24).

Capacities of smooth wall pipe or tubing may also be determined by using the following formulae*:

High Pressure [1.5 psig (10.3 kPa) and above]:

$$Q = 181.6 \sqrt{\frac{D^5 \cdot (P_1^2 - P_2^2) \cdot Y}{Cr \cdot fba \cdot L}}$$

$$= 2237 \, D^{2.623} \left[\frac{(P_1^2 - P_2^2) \cdot Y}{Cr.L}\right]^{0.541}$$

Low Pressure [Less than 1.5 psig (10.3 kPa)]:

$$Q = 187.3 \sqrt{\frac{D^5 \cdot \Delta H}{Cr \cdot fba \cdot L}}$$

$$= 2313 \, D^{2.623} \left(\frac{\Delta H}{Cr.L}\right)^{0.541}$$

where:

Q = Rate, cubic feet per hour at 60°F and 30-inch mercury column.

D = Inside diameter of pipe, inches (mm).

P_1 = Upstream pressure, pounds per square inch.

P_2 = Downstream pressure, psia.

Y = Superexpansibility factor = 1/supercompressibility factor.

Cr = Factor for viscosity, density, and temperature.

$$= 0.00354 \, ST \left(\frac{Z}{S}\right)^{.152}$$

S = Specific gravity of gas at 60°F and 30-inch mercury column.

T = Absolute temperature, °F or = t + 460.

t = Temperature, °F.

Z = Viscosity of gas, centipoise (0.012 for natural gas, 0.008 for propane), or = 1488μ

m = Viscosity, pounds per second ft

fba = Base friction factor for air at 60°F (CF = 1).

L = Length of pipe, feet (m).

ΔH = Pressure drop, in. water column (27.7 in. H_2O = 1 psi).

$$CF = \text{Factor CF} = \left(\frac{fb}{fba}\right)$$

fb = Base friction factor for any fluid at a given temperature, °F.

*For further details on the formulae, refer to "Polyflo Flow Computer," available from Polyflo Company, 3412 High Bluff, Dallas, Texas 75234.

†For values for natural gas, refer to Manual for Determination of Supercompressibility Factors for Natural Gas, available from American Gas Association, 1515 Wilson Boulevard, Arlington, Virginia 22209. For values for liquefied petroleum gases, refer to *Engineering Data Book*, available from Gas Processors Association, 1812 First Place, Tulsa, Oklahoma 74102.

5. Use this vertical column to locate ALL gas demand figures for this particular system of piping.

6. Starting at the most remote outlet, find in the vertical column just selected the gas demand for that outlet. If the exact figure of demand is not shown, choose the next larger figure below in the column.

7. Opposite this demand figure, in the first column at the left, will be found the correct size of gas piping.

8. Proceed in a similar manner for each outlet and each section of gas piping. For each section of piping, determine the total gas demand supplied by that section.

TABLE A-1
APPROXIMATE GAS INPUT FOR
TYPICAL APPLIANCES

APPLIANCE	INPUT BTU PER HR. (APPROX.)
Range, free standing, domestic	65,000
Built-in oven or broiler unit, domestic	25,000
Built-in top unit, domestic	40,000
Water heater, automatic storage 30 to 40 gal. tank	45,000
Water heater, automatic storage 50 gal. tank	55,000
Water heater, automatic instantaneous Capacity 2 gal. per minute	142,800
Water heater, automatic instantaneous Capacity 4 gal. per minute	285,000
Water heater, automatic instantaneous Capacity 6 gal. per minute	428,400
Water heater, domestic, circulating or side-arm	35,000
Refrigerator	3,000
Clothes dryer, type 1 (domestic)	35,000
Gas light	2,500
Incinerator, domestic	35,000

For SI: 1 gallon = 3.785 L, 1 British thermal unit per hour = 02931 W.

For specific appliances or appliances not shown above, the input should be determined from the manufacturer's rating.

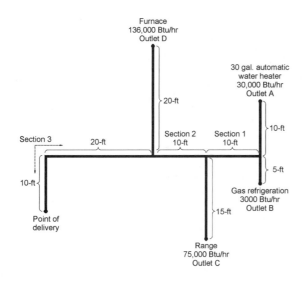

Furnace
136,000 Btu/hr
Outlet D

30 gal. automatic
water heater
30,000 Btu/hr
Outlet A

Section 3   20-ft   Section 2   Section 1
10-ft   10-ft

10-ft   10-ft

5-ft

Point of
delivery

Gas refrigeration
3000 Btu/hr
Outlet B

15-ft

Range
75,000 Btu/hr
Outlet C

For SI: 1 foot = 304.8 mm; 1 gallon = 3.785 L, 1 British thermal unit per hour = 0.2931 W.

EXHIBIT 1

Example of piping system design:

Determine the required pipe size of each section and outlet of the piping system shown in Exhibit 1, with a designated pressure drop of 0.50 inch water column (125 pa). The gas to be used has 0.65 specific gravity and a heating value of 1,000 Btu per cubic foot (37.5 Mj/m^3).

Exhibit 1

Solution:

(1) Maximum gas demand for Outlet A:

$$\frac{\text{Consumption (rating plate input, or Table A-1 if necessary)}}{\text{Btu of gas}} =$$

$$\frac{30,000 \text{ Btu per hour rating}}{1,000 \text{ Btu per cubic foot}} = \frac{30 \text{ cubic feet per hour}}{\text{(or 30 cfh)}}$$

Maximum gas demand for Outlet B:

$$\frac{\text{Consumption}}{\text{Btu of gas}} = \frac{3,000}{1,000} = 3 \text{ cfh}$$

Maximum gas demand for Outlet C:

$$\frac{\text{Consumption}}{\text{Btu of gas}} = \frac{75,000}{1,000} = 75 \text{ cfh}$$

Maximum gas demand for Outlet D:

$$\frac{\text{Consumption}}{\text{Btu of gas}} = \frac{136,000}{1,000} = 136 \text{ cfh}$$

(2) The length of pipe from the point of delivery to the most remote outlet (A) is 60 feet (18.3 m). This is the only distance used.

(3) Using the column marked 60 feet (18.3 m) in Table 402.3(2):

Outlet A, supplying 30 cfh (0.8 m^3/hr), requires ^{3}/$_8$-inch pipe.

Outlet B, supplying 3 cfh (0.08 m^3/hr), requires ^{1}/$_4$-inch pipe.

Section 1, supplying Outlets A and B, or 33 cfh (0.9 m^3/hr), requires ^{3}/$_8$-inch pipe.

Outlet C, supplying 75 cfh (2.1 m^3/hr), requires ^{3}/$_4$-inch pipe.

Section 2, supplying Outlets A, B, and C, or 108 cfh (3.0 m^3/hr), requires ^{3}/$_4$-inch pipe.

Outlet D, supplying 136 cfh (3.8 m^3/hr), requires ^{3}/$_4$-inch pipe.

Section 3, supplying Outlets A, B, and C, or 244 cfh (6.8 m^3/hr), requires 1-inch pipe.

(4) If the gravity factor [*see* (d) under Description of Tables] is applied to this example, the values in the column marked 60 feet (18.3 m) of Table 402.3(2) would be multiplied by the multiplier (0.962) from Table 402.3(13) and the resulting cubic feet per hour values would be used to size the piping.

APPENDIX B
SIZING OF VENTING SYSTEMS SERVING APPLIANCES EQUIPPED WITH DRAFT HOODS, CATEGORY I APPLIANCES, AND APPLIANCES LISTED FOR USE AND TYPE B VENTS

(This appendix is informative and is not part of the code.)

EXAMPLES USING SINGLE APPLIANCE VENTING TABLES

Example 1: Single draft hood-equipped appliance

An installer has a 120,000 Btu per hour input appliance with a 5-inch diameter draft hood outlet that needs to be vented into a 10-foot-high Type B vent system. What size vent should be used assuming (a) a 5-foot lateral single-wall metal vent connector is used with two 90-degree elbows, or (b) a 5-foot lateral single-wall metal vent connector is used with three 90-degree elbows in the vent system?

Solution:

Table 504.2(2) should be used to solve this problem, because single-wall metal vent connectors are being used with a Type B vent.

(a) Read down the first column in Table 504.2(2) until the row associated with a 10-foot height and 5-foot lateral is found. Read across this row until a vent capacity greater than 120,000 Btu per hour is located in the shaded columns labeled "NAT Max" for draft hood-equipped appliances. In this case, a 5-inch diameter vent has a capacity of 122,000 Btu per hour and may be used for this application.

(b) If three 90-degree elbows are used in the vent system, then the maximum vent capacity listed in the tables must be reduced by 10 percent (see Section 504.2.3 for Single Appliance Vents). This implies that the 5-inch diameter vent has an adjusted capacity of only 110,000 Btu per hour. In this case, the vent system must be increased to 6 inches in diameter (see calculations below).

122,000 (.90) = 110,000 for 5-inch vent
From Table 504.2(2), Select 6-inch vent
186,000 (.90) = 167,000; This is greater than the required 120,000. Therefore, use a 6-inch vent and connector where three elbows are used.

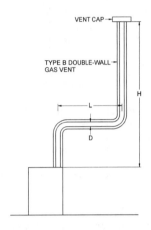

Table 504.2(1) is used when sizing Type B double-wall gas vent connected directly to the appliance.

Note: The appliance may be either Category I draft hood-equipped or fan-assisted type.

FIGURE B-1
TYPE B DOUBLE-WALL VENT SYSTEM SERVING A SINGLE APPLIANCE WITH A TYPE B DOUBLE-WALL VENT

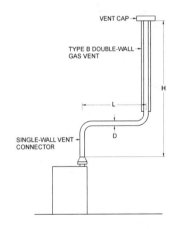

Table 504.2(2) is used when sizing a single-wall metal vent connector attached to a Type B double-wall gas vent.

Note: The appliance may be either Category I draft hood-equipped or fan-assisted type.

FIGURE B-2
TYPE B DOUBLE-WALL VENT SYSTEM SERVING A SINGLE APPLIANCE WITH A SINGLE-WALL METAL VENT CONNECTOR

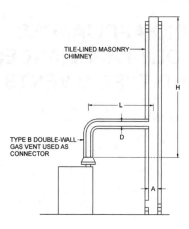

Table 504.2(3) is used when sizing a Type B double-wall gas vent connector attached to a tile-lined masonry chimney.

Note: "A" is the equivalent cross-sectional area of the tile liner.

Note: The appliance may be either Category I draft hood equipped or fan-assisted type.

FIGURE B-3
VENT SYSTEM SERVING A SINGLE APPLIANCE
WITH A MASONRY CHIMNEY OF TYPE B
DOUBLE-WALL VENT CONNECTOR

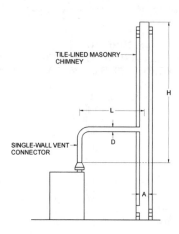

Table 504.2(4) is used when sizing a single-wall vent connector attached to a tile-lined masonry chimney.

Note: "A" is the equivalent cross-sectional area of the tile liner.

Note: The appliance may be either Category I draft hood equipped or fan-assisted type.

FIGURE B-4
VENT SYSTEM SERVING A SINGLE APPLIANCE
USING A MASONRY CHIMNEY AND A
SINGLE-WALL METAL VENT CONNECTOR

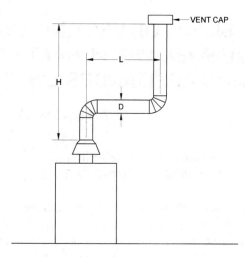

Asbestos cement Type B or single-wall metal vent serving a single draft-hood-equipped appliance [see Table 504.2(5)].

FIGURE B-5
ASBESTOS CEMENT TYPE B OR SINGLE-WALL
METAL VENT SYSTEM SERVING A SINGLE
DRAFT HOOD-EQUIPPED APPLIANCE

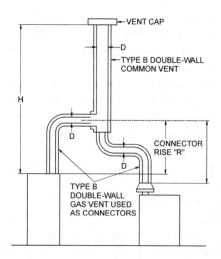

Table 504.3(1) is used when sizing Type B double-wall vent connectors attached to a Type B double-wall common vent.

Note: Each appliance may be either Category I draft hood equipped or fan-assisted type.

FIGURE B-6
VENT SYSTEM SERVING TWO OR MORE APPLIANCES
WITH TYPE B DOUBLE-WALL VENT AND TYPE B
DOUBLE-WALL VENT CONNECTOR

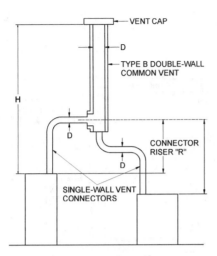

Table 504.3(2) is used when sizing single-wall vent connectors attached to a Type B double-wall common vent.

Note: Each appliance may be either Category I draft hood equipped or fan-assisted type.

FIGURE B-7
VENT SYSTEM SERVING TWO OR MORE APPLIANCES
WITH TYPE B DOUBLE-WALL VENT AND
SINGLE-WALL METAL VENT CONNECTORS

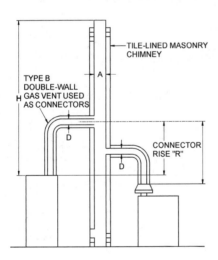

Table 504.3(3) is used when sizing Type B double-wall vent connectors attached to a tile-lined masonry chimney.

Note: "A" is the equivalent cross-sectional area of the tile liner.

Note: Each appliance may be either Category I draft hood equipped or fan-assisted type.

FIGURE B-8
MASONRY CHIMNEY SERVING TWO OR MORE APPLIANCES
WITH TYPE B DOUBLE-WALL VENT CONNECTOR

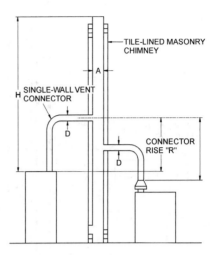

Table 504.3(4) is used when sizing single-wall metal vent connectors attached to a tile-lined masonry chimney.

Note: "A" is the equivalent cross-sectional area of the tile liner.

Note: Each appliance may be either Category I draft hood equipped or fan-assisted type.

FIGURE B-9
MASONRY CHIMNEY SERVING TWO OR MORE APPLIANCES
WITH SINGLE-WALL METAL VENT CONNECTORS

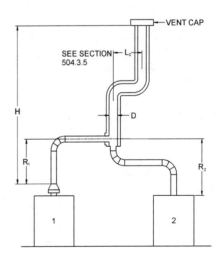

Asbestos cement Type B or single-wall metal pipe vent serving two or more draft-hood-equipped appliances. [See Table 504.3(5)].

FIGURE B-10
ASBESTOS CEMENT TYPE B OR SINGLE-WALL
METAL VENT SYSTEM SERVING TWO OR MORE
DRAFT-HOOD-EQUIPPED APPLIANCES

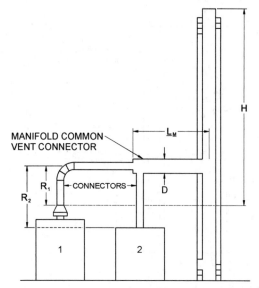

Example: Manifolded Common Vent Connector L_M shall be no greater than 18 times the common vent connector manifold inside diameter; i.e., a 4-inch (1002 mm) inside diameter common vent connector manifold shall not exceed 72 inches (1829 mm) in length (see Section 504.3.4).

Note: This is an illustration of a typical manifolded vent connector. Different appliance, vent connector, or common vent types are possible. Consult Section 502.3.

FIGURE B-11
USE OF MANIFOLD COMMON VENT CONNECTOR

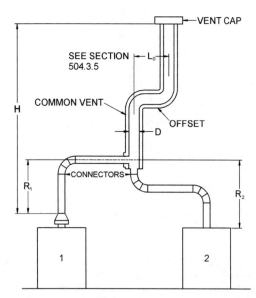

Example: Offset Common Vent

Note: This is an illustration of a typical offset vent. Different appliance, vent connector, or vent types are possible. Consult Sections 504.2 and 504.3.

FIGURE B-12
USE OF OFFSET COMMON VENT

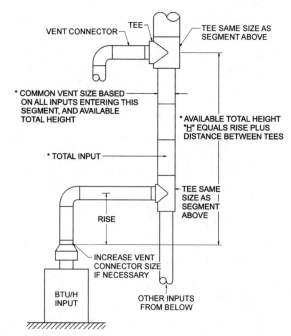

Vent connector size depends on:
- Input
- Rise
- Available total height "H"
- Table 504.3(1) connectors

Common vent size depends on:
- Combined inputs
- Available total height "H"
- Table 504.3(1) common vent

FIGURE B-13
MULTISTORY GAS VENT DESIGN PROCEDURE
FOR EACH SEGMENT OF SYSTEM

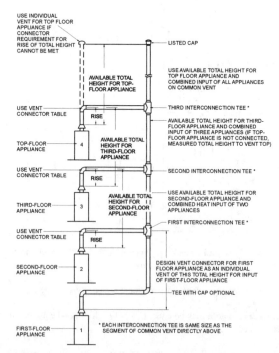

Principles of design of multistory vents using vent connector and common vent design tables (see Sections 504.3.10 through 504.3.15).

FIGURE B-14
MULTISTORY VENT SYSTEMS

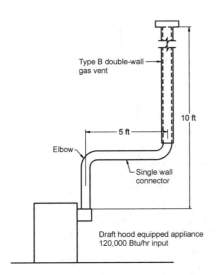

For SI: 1 foot = 304.8 mm, 1 British thermal unit per hour = 0.2931 W.

FIGURE B-15 (Example 1)
SINGLE DRAFT HOOD-EQUIPPED APPLIANCE

Example 2: Single fan-assisted appliance

An installer has an 80,000 Btu per hour input fan-assisted appliance that must be installed using 10 feet of lateral connector attached to a 30-foot high Type B vent. Two 90-degree elbows are needed for the installation. Can a single-wall metal vent connector be used for this application?

Solution:

Table 504.2(2) refers to the use of single-wall metal vent connectors with Type B vent. In the first column, find the row associated with a 30-foot height and a 10-foot lateral. Read across this row, looking at the FAN Min and FAN Max columns, to find that a 3-inch diameter single-wall metal vent connector is not recommended. Moving to the next larger size single wall connector (4 inches), note that a 4-inch diameter single-wall metal connector has a recommended minimum vent capacity of 91,000 Btu per hour and a recommended maximum vent capacity of 144,000 Btu per hour. The 80,000 Btu per hour fan-assisted appliance is outside this range, so the conclusion is that a single-wall metal vent connector cannot be used to vent this appliance using 10 feet of lateral for the connector.

However, if the 80,000 Btu per hour input appliance could be moved to within 5 feet of the vertical vent, then a 4-inch single-wall metal connector could be used to vent the appliance. Table 504.2(2) shows the acceptable range of vent capacities for a 4-inch vent with 5 feet of lateral to be between 72,000 Btu per hour and 157,000 Btu per hour.

If the appliance cannot be moved closer to the vertical vent, then Type B vent could be used as the connector material. In this case, Table 504.2(1) shows that for a 30-foot high vent with 10 feet of lateral, the acceptable range of vent capacities for a 4-inch diameter vent attached to a fan-assisted appliance is between 37,000 Btu per hour and 150,000 Btu per hour.

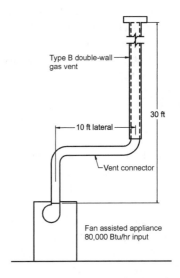

For SI: 1 foot = 304.8 mm, 1 British thermal unit per hour = 0.2931 W.

FIGURE B-16 (Example 2)
SINGLE FAN-ASSISTED APPLIANCE

Example 3: Interpolating between table values

An installer has an 80,000 Btu per hour input appliance with a 4-inch diameter draft hood outlet that needs to be vented into a 12-foot-high Type B vent. The vent connector has a 5-foot lateral length and is also Type B. Can this appliance be vented using a 4-inch diameter vent?

Solution:

Table 504.2(1) is used in the case of an all Type B vent system. However, since there is no entry in Table 504.2(1) for a height of 12 feet, interpolation must be used. Read down the 4-inch diameter NAT Max column to the row associated with 10-foot height and 5-foot lateral to find the capacity value of 77,000 Btu per hour. Read further down to the 15-foot height, 5-foot lateral row to find the capacity value of 87,000 Btu per hour. The difference between the 15-foot height capacity value and the 10-foot height capacity value is 10,000 Btu per hour. The capacity for a vent system with a 12-foot height is equal to the capacity for a 10-foot height plus $2/5$ of the difference between the 10-foot and 15-foot height values, or $77,000 + 2/5 \ (10,000) = 81,000$ Btu per hour. Therefore, a 4-inch-diameter vent may be used in the installation.

EXAMPLES USING COMMON VENTING TABLES

Example 4: Common venting two draft-hood-equipped appliances

A 35,000 Btu per hour water heater is to be common vented with a 150,000 Btu per hour furnace using a common vent with a total height of 30 feet. The connector rise is 2 feet for the water heater with a horizontal length of 4 feet. The connector rise for the furnace is 3 feet with a horizontal length of 8 feet. Assume single-wall metal connectors will be used with Type B vent. What size connectors and combined vent should be used in this installation?

Solution:

Table 504.3(2) should be used to size single-wall metal vent connectors attached to Type B vertical vents. In the vent connector capacity portion of Table 504.3(2), find the row associated with a 30-foot vent height. For a 2-foot rise on the vent connector for the water heater, read the shaded columns for draft-hood-equipped appliances to find that a 3-inch diameter vent connector has a capacity of 37,000 Btu per hour. Therefore, a 3-inch single-wall metal vent connector may be used with the water heater. For a draft-hood-equipped furnace with a 3-foot rise, read across the appropriate row to find that a 5-inch-diameter vent connector has a maximum capacity of 120,000 Btu per hour (which is too small for the furnace) and a 6-inch-diameter vent connector has a maximum vent capacity of 172,000 Btu per hour. Therefore, a 6-inch-diameter vent connector should be used with the 150,000 Btu per hour furnace. Since both vent connector horizontal lengths are less than the maximum lengths listed in Section 504.3.2, the table values may be used without adjustments.

In the common vent capacity portion of Table 504.3(2), find the row associated with a 30-foot vent height and read over to the NAT + NAT portion of the 6-inch diameter column to find a maximum combined capacity of 257,000 Btu per hour. Since the two appliances total only 185,000 Btu per hour, a 6-inch common vent may be used.

Example 5a: Common venting a draft-hood-equipped water heater with a fan-assisted furnace into a Type B vent

In this case, a 35,000 Btu per hour input draft-hood-equipped water heater with a 4-inch-diameter draft hood outlet, 2 feet of connector rise, and 4 feet of horizontal length is to be common vented with a 100,000 Btu per hour fan-assisted furnace with a 4-inch-diameter flue collar, 3 feet of connector rise, and 6 feet of horizontal length. The common vent consists of a 30-foot height of Type B vent. What are the recommended vent diameters for each connector and the common vent? The installer would like to use a single-wall metal vent connector.

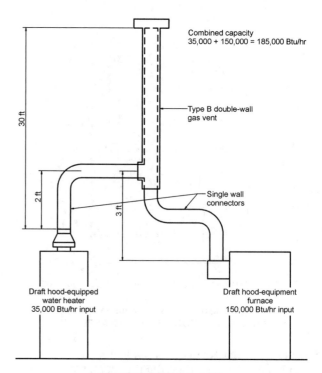

FIGURE B-17 (Example 4)
COMMON VENTING TWO DRAFT-HOOD-EQUIPPED APPLIANCES

Solution: - [Table 504.3(2)]

Water Heater Vent Connector Diameter. Since the water heater vent connector horizontal length of 4 feet is less than the maximum value listed in Section 504.3.2, the venting table values may be used without adjustments. Using the Vent Connector Capacity portion of Table 504.3(2), read down the Total Vent Height (H) column to 30 feet and read across the 2-foot Connector Rise (R) row to the first Btu per hour rating in the NAT Max column that is equal to or greater than the water heater input rating. The table shows that a 3-inch vent connector has a maximum input rating of 37,000 Btu per hour. Although this is greater than the water heater input rating, a 3-inch vent connector is prohibited by Section 504.3.19. A 4-inch vent connector has a maximum input rating of 67,000 Btu per hour and is equal to the draft hood outlet diameter. A 4-inch vent connector is selected. Since the water heater is equipped with a draft hood, there are no minimum input rating restrictions.

Furnace Vent Connector Diameter. Using the Vent Connector Capacity portion of Table 504.3(2), read down the Total Vent Height (H) column to 30 feet and across the 3-foot Connector Rise (R) row. Since the furnace has a fan-assisted combustion system, find the first FAN Max column with a Btu per hour rating greater than the furnace input rating. The 4-inch vent connector has a maximum input rating of 119,000 Btu per hour and a minimum input rating of 85,000 Btu per hour. The 100,000 Btu per hour furnace in this example falls within this range, so a 4-inch connector is adequate. Since the furnace vent connector horizontal length of 6 feet does not exceed the maximum value listed in Section 504.3.2, the venting table values may be used without adjustment. If the furnace had an input rating of 80,000 Btu per hour, then a Type B vent connector [see Table 504.3(1)] would be needed in order to meet the minimum capacity limit.

Common Vent Diameter. The total input to the common vent is 135,000 Btu per hour. Using the Common Vent Capacity portion of Table 504.3(2), read down the Total Vent Height *(H)* column to 30 feet and across this row to find the smallest vent diameter in the FAN + NAT column that has a Btu per hour rating equal to or greater than 135,000 Btu per hour. The 4-inch common vent has a capacity of 132,000 Btu per hour and the 5-inch common vent has a capacity of 202,000 Btu per hour. Therefore, the 5-inch common vent should be used in this example.

Summary. In this example, the installer may use a 4-inch-diameter, single-wall metal vent connector for the water heater and a 4-inch-diameter, single-wall metal vent connector for the furnace. The common vent should be a 5-inch-diameter Type B vent.

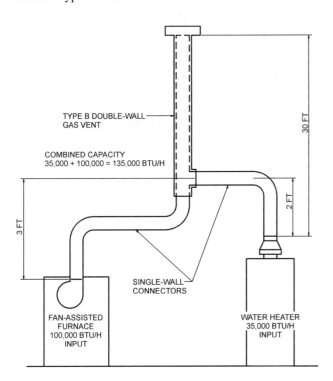

FIGURE B-18 (Example 5a)
COMMON VENTING A DRAFT HOOD WITH A FAN-ASSISTED FURNACE INTO A TYPE B DOUBLE-WALL COMMON VENT

Example 5b: Common venting into a masonry chimney

In this case, the water heater and fan-assisted furnace of Example 5a are to be common vented into a clay tile-lined masonry chimney with a 30-foot height. The chimney is not exposed to the outdoors below the roof line. The internal dimensions of the clay tile liner are nominally 8 inches by 12 inches. Assuming the same vent connector heights, laterals, and materials found in Example 5a, what are the recommended vent connector diameters, and is this an acceptable installation?

Solution:

Table 504.3(4) is used to size common venting installations involving single-wall connectors into masonry chimneys.

Water Heater Vent Connector Diameter. Using Table 504.3(4), Vent Connector Capacity, read down the Total Vent Height *(H)* column to 30 feet, and read across the 2-foot Connector Rise *(R)* row to the first Btu per hour rating in the NAT Max column that is equal to or greater than the water heater input rating. The table shows that a 3-inch vent connector has a maximum input of only 31,000 Btu per hour while a 4-inch vent connector has a maximum input of 57,000 Btu per hour. A 4-inch vent connector must therefore be used.

Furnace Vent Connector Diameter. Using the Vent Connector Capacity portion of Table 504.3(4), read down the Total Vent Height *(H)* column to 30 feet and across the 3-foot Connector Rise *(R)* row. Since the furnace has a fan-assisted combustion system, find the first FAN Max column with a Btu per hour rating greater than the furnace input rating. The 4-inch vent connector has a maximum input rating of 127,000 Btu per hour and a minimum input rating of 95,000 Btu per hour. The 100,000 Btu per hour furnace in this example falls within this range, so a 4-inch connector is adequate.

Masonry Chimney. From Table B-1, the equivalent area for a nominal liner size of 8 inches by 12 inches is 63.6 square inches. Using Table 504.3(4), Common Vent Capacity, read down the FAN + NAT column under the Minimum Internal Area of Chimney value of 63 to the row for 30-foot height to find a capacity value of 739,000 Btu per hour. The combined input rating of the furnace and water heater, 135,000 Btu per hour, is less than the table value, so this is an acceptable installation.

Section 504.3.15 requires the common vent area to be no greater than seven times the smallest listed appliance categorized vent area, flue collar area, or draft hood outlet area. Both appliances in this installation have 4-inch-diameter outlets. From Table B-1, the equivalent area for an inside diameter of 4 inches is 12.2 square inches. Seven times 12.2 equals 85.4, which is greater than 63.6, so this configuration is acceptable.

Example 5c: Common venting into an exterior masonry chimney

In this case, the water heater and fan-assisted furnace of Examples 5a and 5b are to be common vented into an exterior masonry chimney. The chimney height, clay tile liner dimensions, and vent connector heights and laterals are the same as in Example 5b. This system is being installed in Charlotte, North Carolina. Does this exterior masonry chimney need to be relined? If so, what corrugated metallic liner size is recommended? What vent connector diameters are recommended?

Solution:

According to Section 504.3.18, Type B vent connectors are required to be used with exterior masonry chimneys. Use Table 504.3(8) to size FAN+NAT common venting installations involving Type-B double wall connectors into exterior masonry chimneys.

The local 99-percent winter design temperature needed to use Table 504.3(8) can be found in the ASHRAE Handbook of Fundamentals. For Charlotte, North Carolina, this design temperature is 19°F.

Chimney Liner Requirement. As in Example 5b, use the 63 square inch Internal Area columns for this size clay tile liner. Read down the 63 square inch column of Table 504.3(8a) to the 30-foot height row to find that the combined appliance maximum Input is 747,000 Btu per hr. The combined input rating of the appliances in this installation, 135,000 Btu per hr, is less than the maximum value, so this criterion is satisfied. Table 504.3(8b), at a 19°F Design Temperature, and at the same vent height and internal area used above, shows that the minimum allowable input rating of a space-heating appliance is 470,000 Btu per hr. The furnace input rating of 100,000 Btu per hr is less than this minimum value. So this criterion is not satisfied, and an alternative venting design needs to be used, such as a Type B vent shown in Example 5a or a listed chimney liner system shown in the remainder of the example.

According to Section 504.3.17, Table 504.3(1) or 504.3(2) is used for sizing corrugated metallic liners in masonry chimneys, with the maximum common vent capacities reduced by 20 percent. This example will be continued assuming Type B vent connectors.

Water Heater Vent Connector Diameter. Using Table 504.3(1), Vent Connector Capacity, read down the Vent Height (H) column to 30 feet, and read across the 2-foot Connector Rise (R) row to the first Btu per hour rating in the NAT Max column that is equal to or greater than the water heater input rating. The table shows that a 3-inch vent connector has a maximum capacity of 39,000 Btu per hour. So the 35,000 Btu per hour water heater in this example can use a 3-inch connector.

Furnace Vent Connector Diameter. Using Table 504.3(1), Vent Connector Capacity, read down the Vent height (H) column to 30 feet, and read across the 3-foot Connector Rise (R) row to the first Btu per hour rating in the FAN Max column that is equal to or greater than the furnace input rating. The 100,000 Btu per hour furnace in this example falls within this range, so a 4-inch connector is adequate.

Chimney Liner Diameter. The total input to the common vent is 135,000 Btu per hour. Using the Common Vent Capacity Portion of Table 504.3(1), read down the Vent Height (H) column to 30 feet and across this row to find the smallest vent diameter in the FAN+NAT column that has a Btu per hour rating greater than 135,000 Btu per hour. The 4-inch common vent has a capacity of 138,000 Btu per hour. Reducing the maximum capacity by 20 percent (Section 504.3.17) results in a maximum capacity for a 4-inch corrugated liner of 110,000 Btu per hour, less than the total input of 135,000 Btu per hour. So a larger liner is needed. The 5-inch common vent capacity listed in Table 504.3(1) is 210,000 Btu per hour, and after reducing by 20 percent is 168,000 Btu per hour. Therefore, a 5-inch corrugated metal liner should be used in this example.

Single-Wall Connectors. Once it has been established that relining the chimney is necessary, Type B double-wall vent connectors are not specifically required. This example could be redone using Table 504.3(2) for single-wall vent connectors. For this case, the vent connector and liner diameters would be the same as found above with Type B double-wall connectors.

TABLE B-1
MASONRY CHIMNEY LINER DIMENSIONS
WITH CIRCULAR EQUIVALENTS[a]

NOMINAL LINER SIZE (inches)	INSIDE DIMENSIONS OF LINER (inches)	INSIDE DIAMETER OR EQUIVALENT DIAMETER (inches)	EQUIVALENT AREA (square inches)
4 x 8	2-1/2 x 6-1/2	4	12.2
		5	19.6
		6	28.3
		7	38.3
8 x 8	6-3/4 x 6-3/4	7.4	42.7
		8	50.3
8 x 12	6-1/2 x 10-1/2	9	63.6
		10	78.5
12 x 12	9-3/4 x 9-3/4	10.4	83.3
		11	95
12 x 16	9-1/2 x 13-1/2	11.8	107.5
		12	113.0
		14	153.9
16 x 16	13-1/4 x 13-1/4	14.5	162.9
		15	176.7
16 x 20	13 x 17	16.2	206.1
		18	254.4
20 x 20	16-3/4 x 16-3/4	18.2	260.2
		20	314.1
20 x 24	16-1/2 x 20-1/2	20.1	314.2
		22	380.1
24 x 24	20-1/4 x 20-1/4	22.1	380.1
		24	452.3
24 x 28	20-1/4 x 24-1/4	24.1	456.2
28 x 28	24-1/4 x 24-1/4	26.4	543.3
		27	572.5
30 x 30	25-1/2 x 25-1/2	27.9	607
		30	706.8
30 x 36	25-1/2 x 31-1/2	30.9	749.9
		33	855.3
36 x 36	31-1/2 x 31-1/2	34.4	929.4
		36	1017.9

For SI: 1 inch = 25.4 mm, 1 square inch = 645.16 m^2.

a. Where liner sizes differ dimensionally from those shown in Table B-1, equivalent diameters may be determined from published tables for square and rectangular ducts of equivalent carrying capacity or by other engineering methods.

APPENDIX C
RESERVED

APPENDIX D
RECOMMENDED PROCEDURE FOR SAFETY INSPECTION OF AN EXISTING APPLIANCE INSTALLATION

(This appendix is informative and is not part of the code.)

The following procedure is intended as a guide to aid in determining that an appliance is properly installed and is in a safe condition for continuing use.

This procedure is predicated on central furnace and boiler installations, and it should be recognized that generalized procedures cannot anticipate all situations. Accordingly, in some cases, deviation from this procedure is necessary to determine safe operation of the equipment.

(a) This procedure should be performed prior to any attempt at modification of the appliance or of the installation.

(b) If it is determined there is a condition that could result in unsafe operation, the appliance should be shut off and the owner advised of the unsafe condition.

The following steps should be followed in making the safety inspection:

1. Conduct a test for gas leakage.

2. Visually inspect the venting system for proper size and horizontal pitch and determine there is no blockage or restriction, leakage, corrosion, and other deficiencies that could cause an unsafe condition.

3. Shut off all gas to the appliance and shut off any other fuel-gas-burning appliance within the same room. **Use the shutoff valve in the supply line to each appliance.**

4. Inspect burners and crossovers for blockage and corrosion.

5. **Applicable only to furnaces.** Inspect the heat exchanger for cracks, openings, or excessive corrosion.

6. **Applicable only to boilers.** Inspect for evidence of water or combustion product leaks.

7. Insofar as is practical, close all building doors and windows and all doors between the space in which the appliance is located and other spaces of the building. Turn on clothes dryers. Turn on any exhaust fans, such as range hoods and bathroom exhausts, so they will operate at maximum speed. Do not operate a summer exhaust fan. Close fireplace dampers. If, after completing Steps 8 through 13, it is believed sufficient combustion air is not available, refer to Section 304 of this code for guidance.

8. Place the appliance being inspected in operation. **Follow the lighting instructions.** Adjust the thermostat so appliance will operate continuously.

9. Determine that the pilot(s), where provided, is burning properly and that the main burner ignition is satisfactory by interrupting and reestablishing the electrical supply to the appliance in any convenient manner. If the appliance is equipped with a continuous pilot(s), test the pilot safety device(s) to determine if it is operating properly by extinguishing the pilot(s) when the main burner(s) is off and determining, after 3 minutes, that the main burner gas does not flow upon a call for heat. If the appliance is not provided with a pilot(s), test for proper operation of the ignition system in accordance with the appliance manufacturer's lighting and operating instructions.

10. Visually determine that the main burner gas is burning properly (i.e., no floating, lifting, or flashback). Adjust the primary air shutter(s) as required.

    If the appliance is equipped with high and low flame controlling or flame modulation, check for proper main burner operation at low flame.

11. Test for spillage at the draft hood relief opening after 5 minutes of main burner operation. Use a flame of a match or candle or smoke from a cigarette, cigar, or pipe.

12. Turn on all other fuel-gas-burning appliances within the same room so they will operate at their full inputs. **Follow lighting instructions for each appliance.**

13. Repeat Steps 10 and 11 on the appliance being inspected.

14. Return doors, windows, exhaust fans, fireplace dampers, and any other fuel-gas-burning appliance to their previous conditions of use.

15. **Applicable only to furnaces.** Check both the limit control and the fan control for proper operation. Limit control operation can be checked by blocking the circulating air inlet or temporarily disconnecting the electrical supply to the blower motor and determining that the limit control acts to shut off the main burner gas.

16. **Applicable only to boilers.** Determine that the water pumps are in operating condition. Test low water cutoffs, automatic feed controls, pressure and temperature limit controls, and relief valves in accordance with the manufacturer's recommendations to determine that they are in operating condition.

APPENDIX E
STRUCTURAL SAFETY

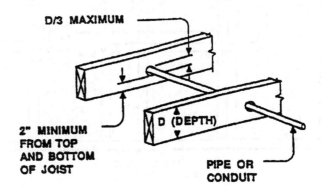

BORED HOLES IN JOISTS

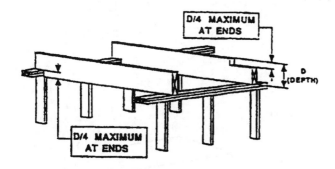

NOTCHES AT ENDS OF JOISTS

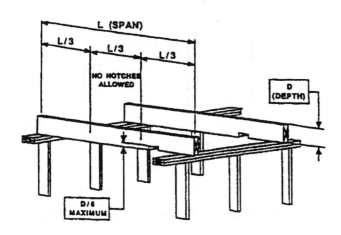

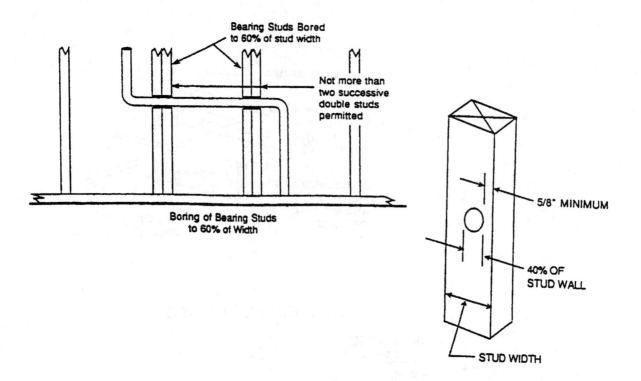

Bearing Studs Bored
to 60% of stud width

Not more than
two successive
double studs
permitted

Boring of Bearing Studs
to 60% of Width

5/8" MINIMUM

40% OF
STUD WALL

STUD WIDTH

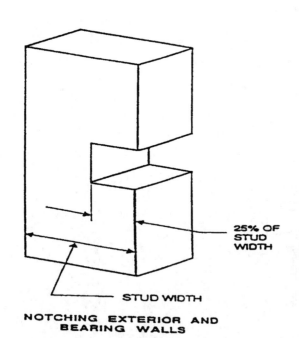

25% OF
STUD
WIDTH

STUD WIDTH

**NOTCHING EXTERIOR AND
BEARING WALLS**

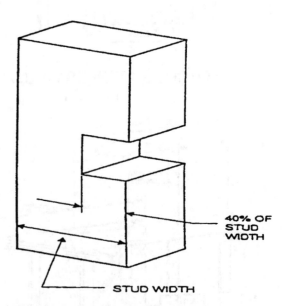

40% OF
STUD
WIDTH

STUD WIDTH

NOTCHING NONBEARING WALLS

FUEL GAS CODE OF NEW YORK STATE

INDEX

A

ACCESS, APPLIANCES
Duct furnaces 609.3
General 306
Shutoff valves 409.1.3, 409.3.1, 409.5
Wall furnaces, vented 607.6
ADJUSTMENTS 607.6, 620.6
ADMINISTRATION ....Chapter 1
Alternate materials and methods 105.2
Alternate methods of sizing
　　chimneys 503.5.5
Appeals 109
Certificates 104.8
Duties and powers of code enforcement official .104
Fees 106.5
Inspections 104.4, 107
Permits 106
Severability 101.5
Scope 101.2
Title 101.1
Violations 108
AIR, COMBUSTION
Requirements ....303.3(1), 304
AIR-CONDITIONING EQUIPMENT 626
Clearances 308.3
**ALTERNATE MATERIALS AND
METHODS** 105.2
APPLIANCES
Broilers for indoor use 622.5
Connections to building piping 411
Cooking 622
Decorative 602
Decorative vented ....202, 303.3(2),
　　　　　　　　　　　　　Table 503.4, 604
Domestic ranges 622.4
Electrical 309
Installation ....Chapter 6
Prohibited locations 303.3
Protection from damage 303.4

B

BENDS, PIPE 405
BOILERS
Existing installations ....Appendix D
Listed 630
Prohibited locations 303.3
Unlisted 631
BUSHINGS ....403.10.4(5), 404.3

C

CENTRAL FURNACES
Existing installation ....Appendix D
CERTIFICATES 104.8
CHIMNEYS ....Chapter 5
Alternate methods of sizing 503.5.5
Clearance reduction 308
Damper opening area 632
Defined 202
CLEARANCE REDUCTION 308
CLEARANCES
Air-conditioning equipment 626.4
Boilers 308.4
Domestic ranges 622.4
Floor furnaces 608.4, 608.6
Open top broiler units 622.5.1
Refrigerators 624.1
Unit heater 619.4
CLOTHES DRYERS
Defined 202
Exhaust 613
General 612
CODE ENFORCEMENT OFFICIAL
Defined 202
Duties and powers 104
COMBUSTION AIR
Confined space 304.10, 304.11
Defined 202
Direct openings outside 304.12.2
Exhaust effect 304.7
Free area 304.10, 304.11,
　　　　　　　　　　　　　304.12, 304.14
Horizontal ducts 304.11.1, 304.12.3
Sources of (from) ....304.15(4)
Sauna heaters 614.5
Unconfined space 304.9, 304.10
Unconfined space, but
　　unusually tight 304.9
Vertical ducts 304.11.1, 304.11.2, 304.12.4
COMPRESSED NATURAL GAS 413
CONCEALED PIPING 404.2
CONDENSATE DISPOSAL 307

CONTROLS
Boilers 630.2
Gas pressure regulators 410.1, 627.4
CONVERSION BURNERS 503.12.1, 618
COOKING APPLIANCES 622
CORROSION PROTECTION 404.8
CREMATORIES 605
CUTTING, NOTCHING, AND
BORED HOLES 302.3, 302.4, 302.5, 302.6

D

DECORATIVE APPLIANCES 602
DEFINITIONS . . . . .Chapter 2
DIRECT VENT APPLIANCES
Defined 202
Installation 304.1
DIVERSITY FACTOR . . . . .Appendix A, 402.2
DRAFT HOODS 202, 503.12
DUCT FURNACES 202, 609

E

ELECTRICAL CONNECTIONS 309.2
EXHAUST SYSTEMS 503.2.1, 503.3.4
F

FEES 106.5
FLOOD HAZARD .301.11
FLOOR FURNACES .608
FUEL CELL POWER PLANTS 633
FURNACES
Central heating, clearance 308.4
Duct 609
Floor 608
Prohibited location 303.3
Vented wall 607

G

GARAGE, INSTALLATION 305.2, 305.3, 305.4
GROUNDING, PIPE 309.1

H

HOT PLATES AND LAUNDRY
STOVES .501.8(3), 622.1

I

ILLUMINATING APPLIANCES 627
INFRARED RADIANT HEATERS 629
INSPECTIONS 104.4, 104.8, 107
INSTALLATION, APPLIANCES
Garage 305

Listed and unlisted appliances 301.3, 305.1
General 301
Specific appliances . . . . .Chapter 6

K

KILNS 628

L

LIQUEFIED PETROLEUM GAS
Defined 202
Motor vehicle dispensing stations 412
Storage 401.2
Systems 402.5.1
Piping material . . . . .403.6.2, 403.11(4)
Thread compounds 403.9.3
Size of pipe or tubing . . . . .Appendix A
LOG LIGHTERS 603

M

MAKE-UP AIR HEATERS 610, 611
Industrial 611
Venting . . . . .501.8(9)
MATERIALS, DEFECTIVE
Repair 301.9
Workmanship and defects 403.7
METERS
Interconnections .401.6
Identification 401.7
Multiple installations 401.6
MINIMUM SAFE PERFORMANCE,
VENT SYSTEMS .503.3

O

OUTLET CLOSURES 404.12
Location 404.13
OXYGEN DEPLETION SYSTEM
Defined 202
Unvented room heaters . . . . .620.6, 303.3(3)

P

PIPE SIZING 402
PIPING
Bends 405
Changes in direction 405
Drips and slopes .408
Installation 404

Inspection 406
Materials 403
Purging 406.7
Sizing 402
Support 407, 415
Testing 406

POOL HEATERS 616

POWERS AND DUTIES OF
THE CODE OFFICIAL .104

PROHIBITED INSTALLATIONS
Floor furnaces 608.2
Fuel-burning appliances 303.3
Piping in partitions .404.2
Plastic piping 404.14.1
Unvented room heater 620.4
Vent connectors . . . . .503.3.3(3)

PURGING 406.7

R

RADIANT HEATERS 629
RANGES, DOMESTIC .622.4
REFRIGERATORS . . . . .501.8(6), 624
REGULATORS, PRESSURE 410.1, 627.4
ROOFTOP INSTALLATIONS 306.5
ROOM HEATERS
Defined 202
Location 303.3
Unvented 620
Vented 621

S

SAFETY SHUTOFF DEVICES
Direct gas-fired make-up
air heaters 610.5
Flame safeguard device 602.2
Unvented room heaters 620.6
SAUNA HEATERS 614
SCOPE 101.2
SEISMIC RESISTANCE 301.12
SERVICE SPACE 306
SPA HEATERS 616
STANDARDS . . . . .Chapter 7
STRUCTURAL SAFETY 302.1
SUPPORTS, PIPING 407, 415

T

TESTING 107
THIMBLE, VENT 503.6.3, 503.10.12,
503.10.16

THREADS
Damaged 403.9.1
Specifications 403.9
TOILETS 625
TRUSS ALTERATIONS 302.7

U

UNCONFINED SPACES 304.10, 304.11
UNIT HEATERS 619
UNLISTED BOILERS .631
UNUSUALLY TIGHT 202, 304.9, 304.12
UNVENTED ROOM HEATERS 620

V

VALVES, MULTIPLE
HOUSE LINES 409.3
VALVES, SHUTOFF
Appliances 409.5
VENTILATING HOODS 503.2.1, 503.3.4
VENTED GAS FIREPLACES
(DECORATIVE APPLIANCES AND HEATERS) . .604
VENTED ROOM HEATERS 621
VENTED WALL FURNACES 607
VENTS
Caps 504.3.14
Direct vent 503.2.3
Equipment not requiring vents 501.8
Gas vent termination 503.6.6
General . . . . .Chapter 5
Integral 505
Listed and labeled 502.1
Mechanical vent 505
VENT, SIZING
Category I appliances 502
Multiappliance 504.3
Multistory 504.3.12, 504.3.13, 504.3.14
Single appliance 504.2
VIBRATION ISOLATION 301.8
VIOLATIONS 108

W

WALL FURNACES, VENTED 607
WARM AIR FURNACES 617
WATER HEATERS 623
Garage installation .305
WIND RESISTANCE .301.10